Integrating English Language Learners in the Science Classroom

Jane Hill
Catherine Little
Jane Sims

Trifolium Books Inc.
A Fitzhenry & Whiteside Company

Trifolium Books Inc.
A Fitzhenry & Whiteside Company
195 Allstate Parkway
Markham, Ontario, Canada   L3R 4T8

In the United States:
Fitzhenry & Whiteside Limited
121 Harvard Avenue, Suite 2
Allston, Massachusetts 02134

Copyright © 2004 Trifolium Books Inc.

All rights reserved. This book is protected by copyright. Permission is hereby granted to the individual purchaser to reproduce the select pages in this book that are so specified for non-commercial individual or classroom use only. Permission is not granted for school-wide, or system-wide, reproduction of materials. No part of this publication may be transmitted, stored, or recorded in any form without the prior written permission of the publishers.

Care has been taken to trace ownership of copyright material contained in this book. The publishers will gladly receive any information that will enable them to rectify any reference or credit line in subsequent editions.

National Library of Canada Cataloguing in Publication
Hill, Jane
  Integrating English language learners in the science classroom / Jane Hill, Catherine Little, Jane Sims.

Includes bibliographical references.
ISBN 1-55244-086-9

1. Science–Study and teaching (Elementary) 2. Science–Study and teaching (Secondary) 3. English language–Study and teaching as a second language (Elementary) 4. English language–Study and teaching as a second language (Secondary) I. Little, Catherine II. Sims, Jane III. Title.

Q181.H54 2004   507.1   C2003-905545-0

Cover design: Kerry Plumley
Cover photo: Jim Cummins/CORBIS/Magmaphoto.com
Page design and layout: Darrell McCalla

Printed and bound in Canada

10 9 8 7 6 5 4 3 2 1

Trifolium's books may be purchased in bulk for educational, business, or promotional use. For information, please write: Special Sales, Trifolium Books Inc. 195 Allstate Parkway, Markham, Ontario, Canada L3R 4T8
Email: info@trifoliumbooks.com

Fitzhenry & Whiteside acknowledges with thanks the Canada Council for the Arts, the Government of Canada through its Book Publishing Industry Development Program (BPIDP), and the Ontario Arts Council for their support in our publishing program.

Acknowledgements

We would like to thank Elizabeth Coelho, Mars Bloch, Ero Siouga and Paula Markus of the Toronto District School Board for bringing us together; Nancy Andraos, our editor for her tact and patience; Brenda Dalgish, Eleanor Minuk, and Norma Rojas for their careful and insightful reviews of this project; the staff and students of St. Timothy Catholic School for help in field testing; the staff of O Heavenly Bagel, the Chinese restaurant for providing endless cups of tea as we argued out the text.

"The Janes" thank Maureen Sims and Bill Oaker for their tolerance of the daily seven a.m. phone calls. Catherine thanks her baby son Alexander who was with us every step of the way and her husband, D'Arcy for his constant encouragement.

Contents

LIST OF HANDOUTS		*v*
INTRODUCTION		*vii*
CHAPTER 1	Teaching Science to English Language Learners	1
	Who Are English Language Learners?	1
	How Long Does it Take to Learn English?	1
	Previous Schooling	2
	Settling-In Pains	2
	Why Science is Difficult for English Language Learners	3
	Strategies for Teaching Beginners	3
	Speaking So English Language Learners Can Understand	4
	Correcting Errors	4
	Use All the Help You Can Get	5
	Fostering Growth	5
CHAPTER 2	Making Lessons Comprehensible	7
	Accessing Prior Knowledge – Forces	7
	Exploring Concepts in First Languages – Cells	12
	Making Lessons Visual, Tactile, and Kinesthetic – Flight	18
	Using Graphic Organizers and Non-Linguistic Representations – Ecosystems	24
	Using Cooperative Learning – Heat	31
CHAPTER 3	Developing Language for Inquiry and Communication	37
	Explaining Vocabulary – Fluids	37
	Practising Asking Questions – Pure Substances And Mixtures	44
	Working with Word Problems – Mechanical Efficiency	47
	Guiding Written Responses – Electricity	52
	Making Textbooks Accessible – Optics	54
CHAPTER 4	Making Connections to the World Outside the Classroom	59
	Implementing Inclusive Curriculum – Space	59
	Using Authentic Reading and Viewing Materials – Water Systems	63
	Constructing Working Models – Motion	66

		Organizing Community Field Trips –	
		Diversity of Living Things	69
		Using the Internet – The Earth's Crust	76
CHAPTER 5		Assessing Fairly	81
		Using Varied Assessment Techniques	81
		Making Accommodations	86

GLOSSARY OF LANGUAGE LEARNING TERMS   89

SUGGESTED WEB SITES   91

RESOURCES AND REFERENCES   95

THE AUTHORS   101

List of Handouts

CHAPTER 2	Handout 2.1: A K.W.L.H. Chart	10
	Handout 2.2: How Many Ways Can You Carry Things	11
	Handout 2.3: A Day in the Life of a Unicellular Organism – Final Copy	15
	Handout 2.4: How to say Predict in Other Languages	16
	Handout 2.5: My Personal Dictionary	17
	Handout 2.6: Lift and Drag	22
	Handout 2.7: What Makes a Kite Fly Well?	23
	Handout 2.8: A Pond Food Web	27
	Handout 2.9: Reading a Food Web	28
	Handout 2.10: A Venn Diagram to Compare Predators and Prey	29
	Handout 2.11: The Effects of Pesticides Flow Chart	30
	Handout 2.12: Notes on Cooking Methods	34
	Handout 2.13: Radiation, Conduction or Convection?	35
CHAPTER 3	Handout 3.1: Word Puzzle on Fluids	40
	Handout 3.2: Clues for Word Puzzle on Fluids	41
	Handout 3.3: Words that Matter Tic-Tac-Toe	43
	Handout 3.4: What if …	46
	Handout 3.5: GRASS (Steps in Solving Word Problems)	50
	Handout 3.6: A Private Survey	51
	Handout 3.7: Steps for Reading	57
CHAPTER 4	Handout 4.1: What Proof Would You Accept?	62
	Handout 4.2: An Interview with an Adult on Water Supplies	65
	Handout 4.3: Planning Sheet for Mechanical Models	68
	Handout 4.4: Permission Letter for Community Field Trips	72
	Handout 4.5: Permission Letter for Community Field Trips *(Spanish version)*	73
	Handout 4.6: Permission Letter for Community Field Trips *(Chinese version)*	74
	Handout 4.7: A Bug's Eye View	75
	Handout 4.8: Making a Classroom Display on Minerals From the Internet	79
CHAPTER 5	Handout 5.1: Checklist of Portfolio Contents	84
	Handout 5.2: Sentence Starters	85

Introduction

Integrating English Language Learners in the Science Classroom has been written for intermediate level teachers (grades 6–8) who have English language learners in their science classroom. There are three main chapters: Making Lessons Comprehensible, Developing Language for Inquiry and Communication, and Making Connections to the World Outside the Classroom. In each of these, we present five strategies for teaching science to English language learners, give specific instructions on how to use the strategy with one of the grade 6–8 science topics and include other ways the same strategy could be used with this topic. The last chapter of the book discusses various ways teachers can assess English language learners.

The activities included in this book have been designed to be used with an entire class. Even first language activities can involve a group of English language learners doing a parallel activity to the rest of the class. Although the material is planned for a classroom science teacher, ESL (English as a Second Language) teachers could use this book to plan content-based lessons. Tutors may also find this material helpful.

The science topics are chosen from the grades 6–8 science curricula currently being implemented in the United States and Canada (Council of Ministers of Education, Canada 1997, and the National Academy of Sciences, United States 1996). Although the material has been prepared for students in grades 6–8, some of the science and all of the teaching strategies are appropriate for students of all ages.

Throughout the book, we hope to dispel the notion that English language learners are more advanced in grade 8 than grade 6. Children in each grade may have no prior English language exposure or may be children with near native English-speaker oral proficiency who are still struggling with academic proficiency in reading and writing. The spiral nature of the science curricula means that topics are explored in different ways at a variety of grade levels.

"I like science, actually our lifes are science. In Korea I studied Science very hard and confused. I used to memorize and didn't try to understand. But now I am trying to understand and it's not very hard."

Science is a wonderful subject for new language students. Even beginners benefit from being integrated into mainstream science classes. Much of the science curricula tend to be international as well as cumulative. Primary classrooms around the world teach about our vast universe. So lessons about the sun warming the Earth, the seasons, and the phases of the moon have formed a foundation for middle school studies about space. The more demanding science curricula at the secondary school level mean that the skills and knowledge taught in middle schools are essential to future understanding and success. Science activities have the potential to be visual, tactile and kinesthetic; students are expected to observe demonstrations, touch materials, and do experiments. The subject is far less alien to English language learners than social studies and literature.

The plethora of acronyms available for referring to students in our classes who are not native speakers of English reflects the diversity of these students and the remarkably wide spectrum of stages language learners pass through, as well as the wish of educators to avoid pejorative labels.

The discomfort with the most commonly used name ESL arises from the inaccuracy of referring to English as a second language when it may well be the learner's third or fourth language. We have chosen to use the practice of the publication *The Science Teacher* in referring to the students as English language learners and the specific language programs that may be available in schools as ESL classes taught by ESL teachers.

Labels for non-English speaking students	
ESL:	English as a Second Language
NEP:	Non-English Speaking
LEP:	Limited English Proficiency
LES:	Limited English Speaking
PEP:	Potentially English Proficient
ELL:	English Language Learners
ESOL:	English to Speakers of Other Languages
EAL:	English as an Additional Language
LM:	Language Minority
CLD:	Culturally and Linguistically Diverse

Teachers of English language learners can dip into this book without necessarily following the content sequentially, whether it is to find an appropriate handout or an engaging class activity. No matter what science topic is being taught in class, teachers can find something useful and ready to use in this book. The material can also be extended to other subject areas. For all of us, diversity is something that enriches all our lives and helps us understand each other better.

Chapter 1

Teaching Science to English Language Learners

English language learners are often marginalized and their opportunities to interact minimalized—even in classrooms of teachers with the best intentions.

(Verplaetse 1998)

Who Are English Language Learners?

We've all experienced the moment when there is a knock on the classroom door at 10:30 a.m. The principal or perhaps a secretary is introducing a new student who has just arrived in this country. But who is this student? She may be a young immigrant who studied the science topic the class is working on two years ago; or a refugee who has spent years in a camp learning only what the adults around him have been able to teach; or perhaps a student whose parents are in North America on business visas who attends weekend and after-school classes to try to keep up with her classmates back home so that she will not have to repeat a grade when she returns.

In other words, English language learners come from very diverse backgrounds and may have different educational needs. English language learners are not a homogeneous group. As teachers, we must develop strategies that will allow us to accommodate the diversity of these students and have them benefit from learning in a mainstream classroom.

How Long Does it Take to Learn English?

Ah, that is the question. These days it seems hard enough to teach science well without the added complication of students who speak little English. In fact, all new students will be in the process of acquiring fluency in academic English, a skill that takes five to seven years to develop.

Some students may, at first, appear not to speak a word of English but already know how to read and write quite well. Others will quickly acquire basic oral skills and chatter quite well in colloquial English. These students, however, may still have serious difficulties with reading and writing. Often we are not aware that these children lack the requisite literacy development because they can effectively participate in oral communication.

Students we label as "weak" may actually be strong enough to no longer require ESL classes but they continue to need your assistance in acquiring academic English. We hope the strategies suggested in this book will be especially helpful to these students. Moreover, by drawing on the contributions from children who have come from different parts of the globe, your science classes will be enriched.

Previous Schooling

The educational background, even of students from the same country, differs widely. Some students from a war-torn country may not read or write in their first language; others may have had access to excellent schools or to private tutoring. Some school systems are free to the end of university; others require parents to pay fees, provide uniforms and transportation and even their own desks. Within the same country, access to schooling may be excellent for citizens but limited for refugees or the children of migrant workers.

A few students may have been educated in a language other than their own. Children of new immigrants to Quebec who later move to English-speaking provinces will likely have had their schooling in French; Afghani children living in Pakistan may have been educated in Urdu rather than in Pushtu.

Settling-In Pains

Concentration can be very difficult for learners of any language. There are so many factors that can affect a language learner's ability to focus on a lesson. English language learners have difficulty identifying our digressions; they may focus unduly on our explanations of the virtues of punctuality. An English word may be distracting, in the way that the Italian word "frigorifero" (meaning "refrigerator") sounds naughty in English. In the struggle to control the giggles that emerge every time such a word or phrase pops up, the English language learner loses the thread of the lesson. Also, listening all day to sounds that don't make much sense is fatiguing, and by the afternoon many of them have tuned out.

Many English language learners fit the stereotype of the eager, hardworking enthusiastic students whose families value education, but others have a very difficult time adjusting to sitting for a long period in a seat. Try to empathize; remember what it was like to try to learn a language or live in a strange city. Imagine having to live through a school day when much of the classroom work is incomprehensible. Don't be discouraged if some days a youngster really doesn't want to stay for extra help.

All English language learners go through the stages of cultural adjustment. At first, new arrivals tend to be fascinated by everything. This euphoric phase may last a few weeks but then culture shock sets in. Some students suffer insomnia; others do nothing but sleep. Some are ravenously hungry and eat too much; others lose their appetites completely and lose weight. Culture shock is often very severe for young teenagers who are just on the edge of puberty. They may be mature enough to see that their parents are struggling and when people at home ask how they are getting along, they answer "fine" even though they may be in crisis. A sympathetic teacher can make a big difference.

Why Science Is Difficult for English Language Learners

Science is a subject that offers students many ways to learn about the topics in the curricula but at the same time there are difficulties. The vocabulary of science is very specialized. Often scientific terms used colloquially do not convey the correct scientific meaning, and many times these terms are used incorrectly. This presents many challenges and can lead to misconceptions for English speakers as well as English language learners.

One textbook may include these sentences in a discussion about osmosis: "The water fills the vacuoles and cytoplasm, causing them to swell up and push against the cell wall. This outward pressure is called turgor pressure." In addition to remembering what vacuoles and cytoplasm are, English language learners must struggle with "swell up." If they check the dictionary they will find many definitions of "swell" to choose from; in this case the addition of the preposition "up" to limit the meaning is not always mentioned. The text may then go on to clarify "turgor pressure" by explaining how road salt causes grass to wilt. An application involving road salt is an unnecessary hurdle for students new to the English language. It is better to provide an example of a wilted plant from a windowsill or do a simple demonstration with celery, salt and fresh water.

Differentiating between technical terms and everyday English, and understanding culture specific examples are only two of the difficulties. English language learners must also master the syntax of comparison and contrast, cause and effect, chronological order, and deductive and inductive reasoning. The discourse of science includes many complex sentences with nuances of meaning based on logical connectors such as "because," "however," "nevertheless," and "if … then."

In addition, the mode of teaching may also be very unfamiliar. Students may come to science class expecting to memorize information and be mystified by teachers who expect them to design their own inquiries and discover the scientific principles themselves. Teachers trying to identify students' misunderstandings find it more difficult to understand how an English language learner may be thinking.

Strategies for Teaching Beginners

There are, however, strategies which will help even English language beginners. This book focuses on intermediate and advanced learners who can speak and write well enough to communicate and understand much of the classroom instruction. When students have very limited English, the key strategy is to let them work along with the rest of the class doing assignments in their first language. This will mean extra work in finding colleagues or senior students to provide indications of how they are doing. However, this is far less time-consuming than trying to dream up language exercises that may pass the time but will not be as productive as encouraging the students to work with the scientific concepts and gradually begin to label diagrams and sketches and fill out charts in English.

One of the stages in language learning is a silent period when learners absorb language and understand far more than they can demonstrate orally. After the students have had time to become acclimatized, try asking questions that have "yes" or "no" answers; then move to questions that require a choice such as: "Did the mercury in the thermometer go up or down?"

Speaking So English Language Learners Can Understand

When you do venture into those first questions, keep in mind that all of us have drifted into speaking more loudly, but it doesn't help. Pausing, not between every other word, but in the natural pauses at the ends of sentences and paragraphs does. Use the active voice rather than the passive because it is clearer. We usually speak in the active voice, not the passive: "Then you put the powder in the test tube" falls from our lips more easily than "The powder is placed in a test tube."

Do keep in mind that English speech is full of idioms that are mystifying to language learners: "You won't get to first base doing it that way"; "Hold your horses"; "Now you're on the right track." The problem is that idioms help the native English-speaking students feel more comfortable whereas English language learners simply do not understand.

Since our vocabulary has many words with Latin roots, speakers of Romance languages such as Italian or Spanish are more likely to be able to understand formal speech than the more colloquial, rather slangy, approach we are accustomed to using when native English speakers don't understand.

Lastly, make your instructions as direct and clear as you can. Minimize the use of digressions and use idiomatic expressions very judiciously. English language learners must spend extra time and effort deciphering digressions and idioms when their attention should really be focused on the lesson itself. Bear in mind that idiomatic expressions take a longer period of time to master than formal speech.

Correcting Errors

Making mistakes is an important part of language learning. Linguists call the error-fraught mélange we utter as we learn a new language "interlanguage". Overt correction of spoken language often does much more harm than good because our well-meaning teaching can result in the student totally clamming up. When a student says: "The volcano he go up high and the lava come down, down," just reply "Yes, you're right. The volcano erupts and then the lava flows down the sides of the volcano."

Respond to content not to form. As a science teacher, you need to be concerned about whether English language learners have understood the science concepts first; the students' ability to express themselves in correct English will come later as they integrate the language with their scientific thought processes.

Set some parameters about writing. If work is to be displayed or published, it should be edited and the student should prepare a corrected copy for a general audience. The best approach to science journal entries is to treat them as dialogue journals and write a response in which you provide some language feedback by modelling correct spelling and syntax. In other work, consult with the ESL teacher so that both of you can concentrate on one language structure at a time. In summative evaluations, it is best to focus only on the content.

Use All the Help You Can Get

In most settings, an ESL teacher will be your main resource. The ideal is to set up a partnership in which the ESL teacher pre-teaches some of the science curriculum so that the students come to class ready to deepen their understanding. Generally, this is not possible because students are often placed based on grade levels rather than according to their language proficiency. However, the study of topics commonly used in ESL classes such as weather, natural disasters, and measurement do include instruction in vocabulary and structures that will assist in understanding science.

If there is no ESL program in your school or if you have only a travelling teacher, take a deep breath. One of the resources described in the Teacher Resources section at the back of this book may help. The most reassuring is *The More-than-Just-Surviving Handbook* by Barbara Law and Mary Eckes (1990). Consider using volunteers to work occasionally with other groups in the class so that you can take a little time with the English language learners yourself.

Fostering Growth

The real art in teaching English language learners is to know when to introduce new challenges. Sometimes both students and teachers become so comfortable with the nurturing approach beginners require, that months later the students are still expecting more assistance than they actually need. It is complicated because sometimes a student may sail through a difficult topic like optics and then be overwhelmed by mechanical efficiency. This may reflect a more thorough background in one topic or the cyclical loss of energy that comes with cultural adjustment.

It is our hope that when you use the strategies in this book you will begin to see a marked increase in the levels of participation of the English language learners in your class, and before long you will also see that they have learned more content than you would have ever thought possible.

Chapter 2
Making Lessons Comprehensible

Accessing Prior Knowledge

The universality of scientific principles, laws and procedures across cultures can help students as they learn about those same principles in a new language. However, effective science learning frequently requires that learners restructure their understandings, change perceptions, and even discard long-held beliefs.

(Fathman, Quinn and Kessler 1992)

The cumulative nature of science curricula means that most English language learners have encountered some aspects of the topic under consideration at an earlier grade level. A little time spent exploring what they already know and bringing the language they need to the surface pays off. English language learners often report that it is only when their teacher is halfway through a topic, such as Forces, that it dawns on them that they have learned something about tension or compression before.

As a science teacher, you are aware of the misconceptions all students, including English language learners, bring to many aspects of science. For example, a student may be convinced that sand is a fluid because it flows. Students from other cultures may have an understanding that you do not anticipate. They may also have studied this topic at an earlier point in their schooling. It is important that they have ample opportunity to explain what they already know.

Some English language learners may have lived in communities less reliant on machines. They may have observed interesting construction methods, various ways of transporting materials and innovative ways of designing toys. Be careful to ensure that students consider experiential learning as well as what they have been taught in school. Integrating the contributions of English language learners will enrich your science classes.

Science is a subject that requires students to build on prior knowledge, both experiential and from previous grades. Make sure you give your students opportunities to share their previous knowledge with you. You may need to modify a lesson or add an activity to address misconceptions they may have about a particular topic. Spending extra time at this stage of their learning is well worth it, as it gives students an opportunity to challenge previously held ideas and helps build a solid foundation for future learning.

Try the timed science journal writing activity on the next page Merchant and Young call "Corners" (Irujo 2000).

"I used to think there was a long electrical cord from the toaster to Niagara Falls."

Activity

Tension Forces in Everyday Life

Materials
3 elastic bands of different thicknesses and lengths
a model or a picture of an interesting suspension bridge
a question card

- Create 3 stations or corners in the room. In each, place a question or an object. For the topic of tension, these might be a display of different elastic bands with instructions to make a list of the best uses of each, a model of a bridge with instructions to sketch the bridge using arrows to show the points of tension and a question card asking "Where is tension used in everyday life?"

- Divide the class into 3 groups.

- Have each group write in their science journals for 10 minutes at one of the three corners and then proceed to the next. Beginners may use their first language.

- As a class, discuss the students' ideas and any difficulties they had.

- Respond to the journal entries with personal notes.

An example of a student answer to corner one:

There were three elastic bands on the table. One was thin and small. One was fat and medium length. The third was thin and long. My sister uses a small thin one to put her hair together. I saw a fat, medium one around letters the postman left. I don't know what to use the thin long one for.

An example of a teacher response to the student:

I use a small, thin elastic to put my hair into a pony tail in the summer. Why do you think it is better than using a fat elastic or a very long elastic? Would it be a good idea for the postman to use a thin elastic? When you have thought about these things, try to think of a way to use a long, thin elastic even if you haven't seen it used before.

Other Ways to Access Prior Knowledge

By refusing to treat second language learners as blank slates and by constantly seeking ways of drawing on their prior knowledge, learning is made more relevant for bilingual children and much richer for their monolingual peers.

(Edwards 1998)

☞ Use charts, such as *Handout 2.1*, with the headings "What We Know," "What We Want to Learn," "What We Learned," and "How We Learned" to help students step through their thinking processes. These charts, also known as K.W.L.H. charts, can be used with any science topic you may be teaching with your class. Below is an example of a K.W.L.H. chart that has been partially completed for Forces.

What We Know	What We Want to Learn	What We Learned	How We Learned
• Forces can be large or small • Forces can make things hard to do • Things break because of forces.	• How can I create a large force? • How can forces help me do things?		

***Fig. 2.1** Forces K.W.L.H. Chart. Students should start the chart at the beginning of a topic and add to it throughout the course of study. The first two columns can be completed first while the last two are ongoing.*

☞ Science journals provide students with opportunities to express their knowledge about science and monitor their own progress in the subject. English language learners can find these especially useful, both to practise their English writing skills and/or express themselves using visual devices. Use journal starters such as these:
- The heaviest thing I've ever moved …
- If I had to lift an elephant, I would …
- The most creative way I've moved something was …

☞ Encourage students to make diagrams and sketches in their journals and remember to respond to content rather than form.

☞ Group brainstorming is a way to lower the anxiety of English language learners about responding in front of the whole class. Have groups of 3–4 students follow the directions on *Handout 2.2: How Many Ways Can You Carry Things?* Since the responses are sketches, English language learners can be recorders. Then create a master list by asking a reporter from each group to give one idea.

A K.W.L.H. Chart

What We Know	What We Want to Learn	What We Learned	How We Learned

How Many Ways Can You Carry Things?

- Work with your group to suggest 6 more ways to carry things. Have your recorder sketch your suggestions. Note: Your suggestions cannot use electricity, power or gas.

- Remember the rules of brainstorming: Accept all other ideas. Never be rude about your group members' ideas.

- Your time limit is 10 minutes.

Exploring Concepts in First Languages

An insistence on 'English Only' may limit students' cognitive activity to their level of proficiency in their second language.

(Ontario Ministry of Education and Training 1999).

In the past, you may have discouraged your students from ever using their first languages. However, a deliberate and careful use of first language helps students to explain what they know about a concept more fully than they would be able to in English. Sometimes it is surprising to note both their knowledge of science and their high levels of literacy. Unless students have access to working with higher level concepts in first language, their proficiency in science is limited to their proficiency in English.

Figures 2.2 and *2.3* illustrate the strategy of using a student's first language by using a piece of creative writing on unicellular organisms. Since storytelling is universal, students are familiar with narrative prose and enjoy using their imaginations.

하나는 동물성 유글레나 다른하나는 식물쪽으로 된 유글레나. 하지만 그들은 언제나 견원지간이었다. 동물쪽인 유글레나는 음식을 얻으려면 사냥을 해야한다고 몸을 가만히 있지않는 반면에 식물쪽인 유글레나는 자신이 엽록소를 가지고 있기때문에 자신은 식물처럼 태양 빛만 받고도 살수있다고 했다. 그래서 그들은 각자대로 나뉘어져 살기로 하였다.

Fig. 2.2 *Peter, who emigrated from Korea four months ago, is writing a story about how the two aspects of the euglena are combined in a unicellular organism.*

one eugelena looks like Animal cells, an another eugelena looks like plant cells, so their living method is very different and they fight everyday. Animal eugelena want go out and hunting food but Plant eugelena want to stay in home and make food. It has chloroplasts, so it can make food from sunlight. But Animal eugela and plant eugela fight everyday.

Fig. 2.3 *The English translation is richer in content than would have been possible if Peter had written it in English directly.*

Life as a Unicellular Organism

Materials
Handout 2.3: A Day in the Life of a Unicellular Organism—Final Copy

- Have students imagine what life would be like if they were a unicellular organism.

- Then tell students to write a rough draft of what they imagined using their first language.

- Arrange writing conferences with English language learners who have not written in English.

- As each student is explaining the written work, jot down key words the student is searching for.

- Now ask the students to rewrite the story in English using the word list you created and *Handout 2.3: A Day in the Life of a Unicellular Organism—Final Copy*.

Community Connection

The web site **www.yourdictionary.com/languages** provides dictionaries and wordlists in 270 languages. However, if the language is not written in Roman script, specific software may be required. Local communities are likely to be able to help

Community Connection

Web sites often use computer translation programs such as Babel Fish (**see babelfish.altavista.com**). Ask parents and heritage language teachers to help students sort out the peculiar language that results. In the process, they will also build background knowledge in science.

Other Ways to Explore Concepts in First Languages

The point of note taking is to record and remember information; the choice of language is usually irrelevant. The same is true of a number of other activities including brainstorming, planning and discussion within same language groups, where the focus is on developing ideas rather than communicating with the wider group.

(Edwards 1998)

☞ Display a multilingual poster of key scientific thinking skills such as predicting see *Handout 2.4: How to Say Predict in Other Languages*. Such posters can also be made for "analyze" or "conclude."

☞ Have English language learners keep a bilingual personal dictionary as one section of their science notes using *Handout 2.5: My Personal Dictionary* as a model.

☞ When you show your science class a video, encourage English language learners to make a few notes in their first language.

El Núcleo y el Nucleolus

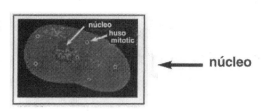

El núcleo y el Nucleolus [núcleo]. El núcleo es el organelle más obvio de cualquier célula eucaryotic. Es membrana-limita el organelle y es rodeado por una membrana doble. Se comunica con el cytosol circundante via poros nucleares numerosos.[huso mitotic]

Dentro del núcleo está la DNA responsable de proveer de la célula sus características únicas. La DNA es similar en cada célula del cuerpo, pero dependiendo del tipo específico de la célula, algunos genes se pueden dar vuelta encendido o apagado – que es porqué una célula del higado es diferente de una célula del músculo, y una célula del músculo es diferente de una célula gorda. Cuando una célula se está dividiendo, la DNA y la proteína circundante condensan en cromosomas (véase la foto) que sea visible por microscopia.

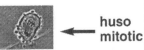

La estructura prominente en el núcleo es el nucleolus. El nucleolus produce los ribosomes, que se mueven del núcleo a las posiciones respecto al retículo endoplasmic áspero donde están críticos en síntesis dela proteína.

Fig. 2.4 *Information in first language helps student comprehension*

Use multilingual web sites such as **www.cellsalive.com**. The internet is a convenient resource that offers a wealth of educational web sites that feature science topics. But beware; surfing the internet can waste enormous amounts of time, so develop a list of web sites from reliable sources for your reference. Make sure that you preview the sites to make sure that the material is appropriate for your students.

A Day in the Life of a
Unicellular Organism—Final Copy

- Imagine what life would be like if you were a unicellular organism.
- Write a rough draft of what you imagined. Don't forget you are responsible for **feeding, digestion, excretion,** and **reproduction.**
- Write your final version on this sheet. You can use the words in your word list to help you.

Hello. I am a _____. This morning I …

How to Say Predict in Other Languages

ਭੱਵਿੱਖ-ਬਾਣੀ
Punjabi

預測
Cantonese

ΜΑΝΤΕΨΕ, ΥΠΟΘΕΣΕ
Greek

அனுமானம்.
Tamil

Predvideti
Serbian

Ka feker
Somali

推测
Mandarin

비가 올 예정입니다.
Korean

تَنَبَّأ
خَمِّن
Arabic

PRZEWIDZIEĆ
Polish

PREdecir
Spanish

આગાહી રાખો.
Gujarati

පුරෝකථනය කරනවා.
Sinhala

My Personal Dictionary

- Write any English words that you are not sure about in the first column.
- What do those words mean in your first language? Use a dictionary or ask your science or ESL teacher.
- Draw a diagram if it makes it easier for you to remember.

English Word	My Language or Sketch
1.	
2.	
3.	
4.	
5.	
6.	
7.	
8.	
9.	
10.	
11.	
12.	

Making Lessons Visual, Tactile, and Kinesthetic

Visuals, objects and living things that can be touched and manipulated, help in making the connections between words and meanings that are needed in order for understanding to occur.

(Fathman, Quinn and Kessler 1992)

English language learners should be doers not just watchers. Concrete experiences lead to the understanding of abstract concepts. Language acquisition is facilitated by movement. Most initial language teaching makes some use of total physical response (TPR). This methodology consists of learners demonstrating their understanding by obeying commands. It is important that English language learners have opportunities to see, touch and manipulate. Such strategies as passing around examples of materials to be used, showing instead of telling or doing an extra experiment are helpful.

In science class, students can demonstrate their reading comprehension without using words if you place materials to be used on a table, give them a list and ask them to assemble the supplies for their group. You can observe their ability to follow complicated oral instructions such as: "Fold your paper lengthwise into four sections." As language learning progresses, students can begin to orally describe what they are doing as they manipulate materials themselves.

An activity that provides both tactile and kinesthetic teaching opportunities is kite-flying.

The Science of Making and Flying Kites

Materials
scraps of paper
1 or 2 kites
Handout 2.6: Lift and Drag
Handout 2.7: What Makes a Kite Fly Well?

- Begin with an active survey by posing three questions:
  How many times have you flown in an airplane?
  Have you ever dreamt you were flying?
  Have you ever flown a kite?

- Have students write answers to these questions on scraps of paper of various dimensions and weights. Then invite them to shape their answer sheets into darts or airplanes.

- Have groups of students throw them one at a time. Discuss why some are more successful than others.

- Then have a group of students gather the papers, and for homework, sort out the answers on chart paper and present the results on the board as a bar graph.

- Obtain a kite or two.

- Wait for a suitable day for flying a kite. The ideal is a day when the weather is warm, a little humid with partly cloudy skies. On such a day there is a good chance of taking advantage of thermals, rising bodies of warm air.

- Take the class to a flying area in the school yard or a nearby park or beach away from power lines, telephone wires and kite-eating trees.

- Have experienced students launch your kite.

- When the kite is aloft, have each student hold the line and make observations of what they feel.

- Have the students who are watching observe the kite's movement and note what factors contribute to successful flight.

- Back in the classroom, use *Handout 2.6: Lift and Drag* to explain how angling the kite takes advantage of the wind's force to provide lift.

- Distribute *Handout 2.7: What Makes a Kite Fly Well?* Ask groups of students to reach a consensus about what the three most important factors are in making a kite fly.

- Explain that scientists call these factors "variables."

- Have students write a summary of the variables that affect flight using *Handout 2.7* as a scaffolding.

Community Connection

Enlist families to help with both construction and kite-flying.

Other Ways to Make Lessons Visual, Tactile, and Kinesthetic

Diversity is a resource and not a problem.

(Edwards 1998)

☞ Consider a cross-curricular project on kites. The design elements could be taught as mathematics, the decoration as art, the library research as language, and the aerodynamics of flight as science. Consider which materials will be readily available to students in their homes and communities and provide more difficult-to-obtain materials such as balsa wood and tissue paper. Make sure students choose designs of kites that actually fly; many cultures have ceremonial kites that are not functional.

☞ Provide a group of students with an imaginary budget of $100. Send them shopping for you in the Toys for All Ages aisle at the Museum store of the National Air and Space Museum at **www.smithsonianstore.com**. Ask them to choose items that would help you teach about the science of flight and justify their choices.

☞ Conduct a "gallery walk" by visiting the How Do Things Fly gallery of the National Air and Space Museum at **www.nasm.edu**. Divide the students into groups to try out the dozens of hands-on activities. Have them explain how the activities work to each other.

☞ Link science curriculum to language arts with a science journal entry modelled on this passage:

> **"The Day We Flew Kites"**
>
> *Flying a kite is pure joy. Once it is up in the air and hanging in the wind, your imagination hangs you up there with the kite. You are the one with the long tail flapping with every gust and change of direction. Your skin is stretched taut to catch the air that supports you above the ground. Only your eyes remain on the ground. They are there to take care of the string that keeps you safe. If that string snaps or escapes, there is danger. There is always a little fear. But why fly if there isn't a bit of risk?*

☞ Have students write a five line poem, about how it feels to fly. Each line in the poem corresponds to 1 syllable, 3 syllables, 5 syllables, 3 syllables and 1 syllable respectively. An example is given on the next page

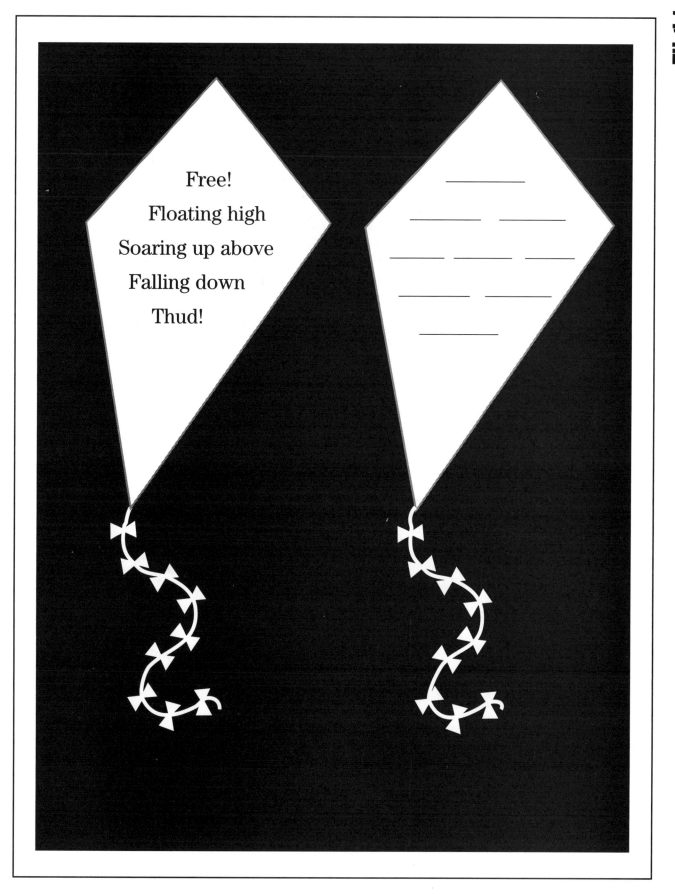

Lift and Drag

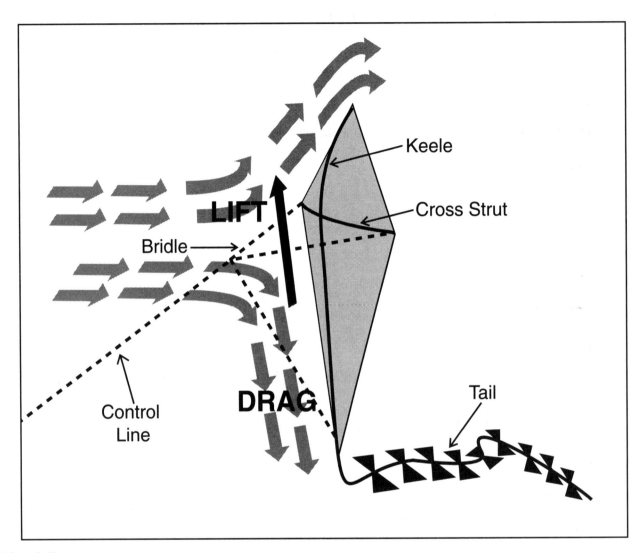

Fig. 2.5 *Lift and Drag.*

What Makes a Kite Fly Well?

- All of the variables listed in the charts below affect how a kite flies.
- With a partner agree on which three are the most important and which three are the least important.
- What observations did you make during the kite-flying activity? Write them beside the appropriate variables.

Properties of a Kite	Observation Made When Flying a Kite
Mass	
Size	
Size of tail	
Balance	
Depth of bow	
Bridling	

Properties of the Wind	Observation Made When Flying a Kite
Speed	
Steadiness	
Air pressure created by the wind	
Altitude	

Other Factors	Observation Made When Flying a Kite
The land—flat or hilly	
Permeability of kite material – paper or cloth	
Rigidity of form	

Using Graphic Organizers and Non-Linguistic Representations

Comprehension can be facilitated through the use of photographs, illustrations, maps, graphs, diagrams and other graphic organizers such as Venn diagrams, semantic webs, and time lines. This kind of scaffolding enables ESL students to participate effectively in instruction even when their knowledge of the language is still quite limited.

(Cummins 1997).

Graphic organizers are diagrams designed to help students see relationships. Because of the built-in focus on higher order thinking skills, the use of such diagrams is especially useful for English language learners. Flow charts establish a sequence of events and thus highlight cause and effect relationships; Venn diagrams help students see similarities and differences; concept maps can be used to draw out prior knowledge and experience, and to summarize what students have learned. Once completed, an organizer becomes an excellent scaffold for writing.

Graphic organizers such as food webs provide English language learners with less language. The convention of arrows pointing to the mouth of the predator to visually describes a concept. A way to help students comprehend the complexity of food webs is to have them construct non-linguistic representations of food chains using *Handout 2.8: A Pond Food Web*.

Activity

The Dynamics in a Food Web

Materials
Handout 2.8: A Pond Food Web
Handout 2.9: Reading a Food Web
a mobile of another food chain
chart paper
construction paper, string, glue, tape

Use the activity "The Dynamics in a Food Web" as an opportunity to give a vivid illustration of the differences between the active and passive voice of the verb: to eat vs to be eaten.

- Explain to the class that they will examine a food web and study it to identify various food chains.

- As a first step, students need to examine the food web in a class discussion. *Handout 2.8: A Pond Food Web* shows a food web which may exist in a pond.

- Define a food chain and list examples on the board. Point out that a food web consists of many food chains.

Fig. 2.6 **Diagram of a Food Chain Mobile.** *The food chain mobile clearly depicts which animal is eaten by another.*

- Have students complete *Handout 2.9: Reading a Food Web.*

- Show your students an example of a mobile from another food chain such as Fig. 2.6 and explain that they will be making their own mobiles.

- Begin by having pairs of students identify and illustrate food chains on chart paper.

- Then have teams plan the size of each animal involved in the food chain.

- Provide materials such as construction paper, string, glue and tape and ask students to make their mobiles.

- As a final step, have students use their work to write a science note that begins: *In a pond, many organisms live side by side. The ... consumes ... for food and is eaten by....*

Other Ways to Use Graphic Organizers and Non-Linguistic Representations

The interrelationships between graphic and linguistic realizations of meaning (as well as the interrelationships between the linguistic modes) can be exploited to make communication clearer and lower the language barrier for students who are learning subject matter knowledge in a second language.

(Early 1990)

☞ Summarize a class discussion on the differences between predators and prey with a Venn diagram using *Handout 2.10: A Venn Diagram to Compare Predators and Prey*. Remind students that only items that are common to both predators and prey should be written in the area where the circles overlap; any other items should be written in the remaining part of the appropriate circles. Provide examples of the uses of connectors for comparing (both … and, neither … nor) and for contrast (however, although).

☞ Textbooks often use pyramids to depict the trophic levels (theoretical levels within an ecosystem depicting the transfer of matter and energy as caused by grazing, predation, parasitism, decomposition, etc.) of a food chain: primary producers, herbivores, different levels of carnivores and decomposers. Explain explicitly why each organism is placed on each trophic level. Explain that the shape of the pyramid has significance; it depicts the relative numbers of organisms within a particular trophic level. The organisms found at the bottom of the pyramid exist in great numbers while the ones at the top occur less often.

☞ Use the cause and effect organizer *Handout 2.11: The Effects of Pesticides Flow Chart* to help students understand how human actions can disrupt an ecosystem. Have students, working in pairs, form sentences using the word "because." Then have them write a summary of the flow chart using these sentences as a scaffolding. You can take this activity a step further by asking the students what we as individuals can do about pesticides.

Community Connection

Have your class survey their families about pest control in both international and North American farms and gardens.

A Pond Food Web

Identify as many food chains as you can in this food web.
- Note how real the animals look.
- Magnification bubbles are used for very small species.
- Remember: The arrow always goes into the mouth of the animal that is eating.

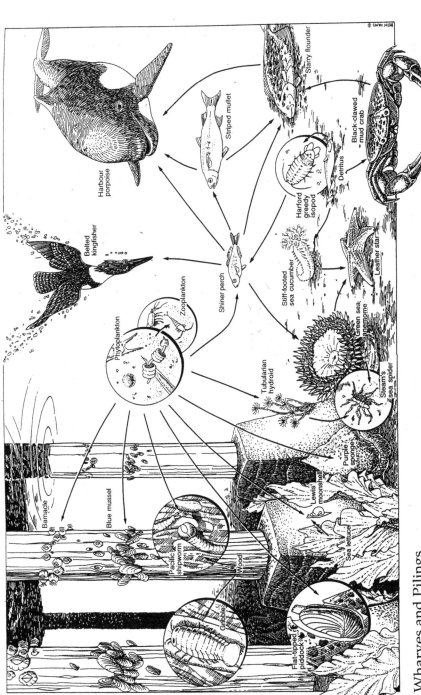

Fig. 2.7 Wharves and Pilings
Protected marine environment, Pacific Coast

Reading a Food Web

Use these questions to fill in the chart.
- Under "Species" what are the names of the living things in the Pond Food Web handout.
- Under "Diet" what does each organism eat?
- Under "Predator" who or what eats that organism?
- Under "Notes" do you notice anything interesting about how the organisms in the food chain are related?

Species	Diet	Predator	Notes

Permission is granted to reproduce this page for purchaser's class only. Copyright © Trifolium Books Inc.

A Venn Diagram to Compare Predators and Prey

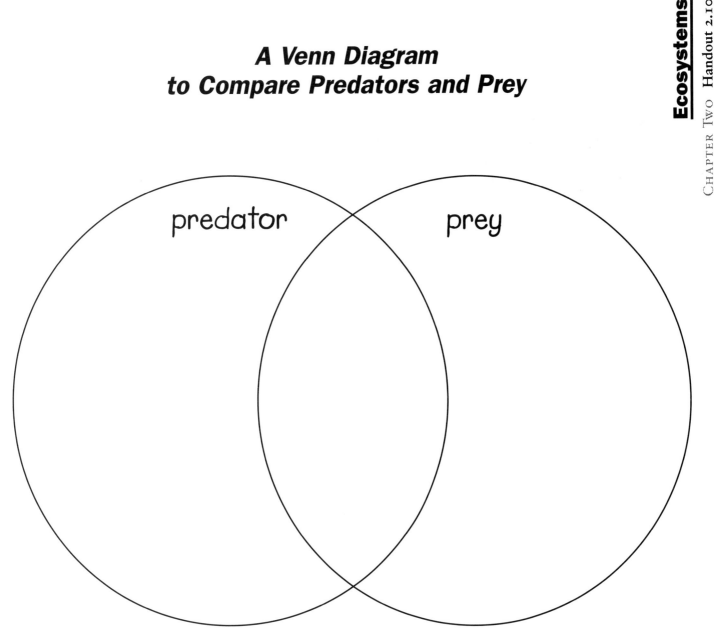

On Venn diagram:
What characteristics apply to:
- both predators and prey? Write this in the area where the circles overlap.
- predators only? Write this in the rest of the circle labelled "predator."
- prey only? Write this in the rest of the circle labelled "prey."

In a note:
Write sentences that describe the ideas shown in the Venn diagram.
- When two things are similar, use a sentence that begins with "both" and continues with "and."
- When things are different, begin a sentence with "although" or use "however" as a joining word.

The Effects of Pesticides Flow Chart

- As a class, use the diagram to talk about how pesticides affect the environment.
- What can we, as individuals and as a community, do about pesticide use?
- Then individually, write a paragraph summarizing the discussion.

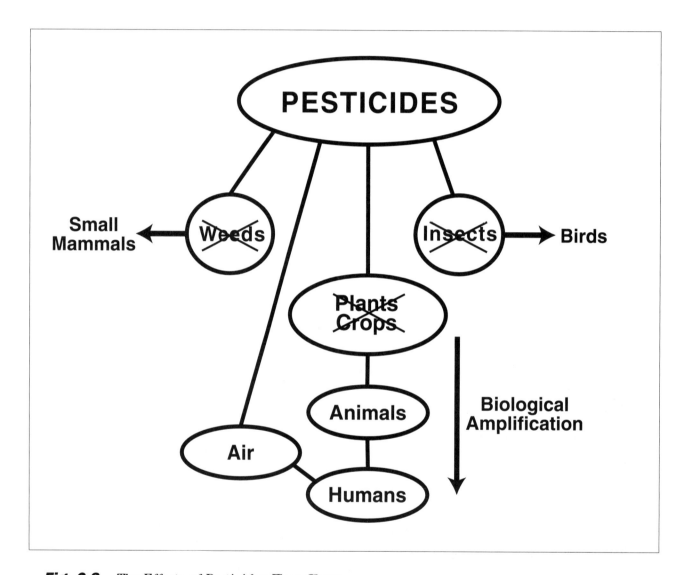

Fig. 2.8 *The Effects of Pesticides Flow Chart*

Using Cooperative Learning

Nobel prizes in science—great breakthroughs in understanding the wonders of the universe—are almost always the result of cooperative learning. Working together, sometimes across oceans and continents, often through using English as a second language, scientists learn from one another, and together make discoveries that ultimately benefit us all.

(Fathman, Quinn and Kessler 1992).

Despite the stereotype of "the mad scientist" working alone in a laboratory, scientific problem-solving is based on teamwork. Traditionally science teachers have grouped students for practical work because of limitations of materials and space. This practice can be an opportunity for cooperative learning.

It is important for English language learners to be included in these groups because rich opportunities for language learning occur in group activities: information must be requested and given, opinions discussed, findings noted, and reports written. Being able to understand or make a small contribution with a group of peers is very affirming for English language learners.

However, just putting the class in groups and telling them to carry on does not ensure cooperative learning is taking place. Lab groups need definite tasks. Post a list such as the one shown below and rotate the tasks weekly.

Week of October 6

Gatherer: Gets materials needed for activity—Josh, Abdul, Rosa, Alysha

Cleaner: Puts away, cleans materials at the end of each activity—Jennifer, Maria, Michael, Myron

Work Monitor: Makes sure each person records necessary data, collects homework, hands out paperwork—Shahzad, Meredith, Pedram, Suzanna

Schedule Monitor: Checks and signs assignment notebook, notes if a group member is absent and saves work for that person, keeps track of time—Jorge, Jonah, Alejandra, Sanaz]

Fig. 2.8 *Examples of Categories from Merchant and Young. (Irujo 2000)*

It is important that groups are flexible and that students understand that they will eventually work with each member of the class. Absolutely random groups formed by choosing a number, a playing card or a coloured marker are good for introducing students to a variety of partners and work best for brief tasks. Sometimes when concepts are difficult or prior knowledge is being explored, first language groups are best. Most of the time, heterogeneous groups are the ideal. Often the group dynamics in a middle school class, the way students in a group interact and behave with each other, will dictate who is able to work together.

Cooperative learning is useful when doing research. An activity such as the gallery walk in "Evaluation of Different Cooking Methods" gives students a chance to hone their presentation skills without having to embarrass themselves in front of the class.

Activity

Evaluation of Different Cooking Methods

Materials
Handout 2.12: Notes on Cooking Methods
materials to make posters
Note: students must do research for this activity.

- As a class, brainstorm various methods of cooking. Introduce the idea of solar cookers and using biomass as an energy source if students do not know about these.

- Divide the class into three groups.

- Have each group of students research the advantages and disadvantages of one of the three most interesting methods discussed. Use *Handout 2.12: Notes on Cooking Methods*.

- Have students complete their third of the handout and make individual posters based on these notes describing the cooking methods, and the advantages and disadvantages of the method.

- When the posters are completed, use an adjacent hallway as an art gallery and hang the posters according to topic.

- Have each group in turn be available for interviews with their classmates to explain their posters and help their classmates complete the other two sections of their handouts.

- Check the completed handouts and have students use them as information for a science note on this application of heat.

Help students recognize key words like these they may come across in their reading.

Plus and Minus Words

Plus (+)	Minus (−)
pro	con
for	against
benefit	risk
positive	negative
advantage	disadvantage

Other Ways to Use Cooperative Learning

Using a collaborative group structure, teachers encourage interdependency among group members, assisting students to work together in small groups so that all participate in sharing data and in developing group reports.... The teacher's role in these small and larger group interactions is to listen, encourage broad participation, and judge how to guide discussion—determining ideas to follow, ideas to question, information to provide, and connections to make.

(National Academy of Sciences 1996)

There are many variations of cooperative learning that you can use in the classroom. You can change the number of students grouped, the way individuals within groups share and the level of competition, if any, between groups.

- Think-Pair-Share: Organize students in pairs. Set a task that requires some thought. For example, list all the ways to heat a house; decide which way is the best. Have students talk about their answers and then share with the class.

- Heads Together: Make an overhead transparency of *Handout 2.13: Radiation, Conduction or Convection?* Place students in teams, randomly or according to group dynamics. Have them number off. Reveal the examples on the transparency one by one, asking which mode of heat transfer is illustrated. Students put their heads together and make sure everyone on the team knows the answer. Call a number and the students raise their hands to answer.

- Jigsaw Reading: Locate a reading such as a newspaper or magazine article. Cut it into 5 sections. Prepare a question sheet with 10 questions distributed throughout the reading. Give groups of students the whole question sheet and one of the sections. Set a time limit. Have the students work together in these expert groups to determine which questions they can answer from their reading. Have them find the answers and rehearse explaining them. Reconfigure the groups to include a member from each expert group and have them complete the answers on the question sheets.

Community Connection

Have students ask their grandparents how they kept cool in the summer when they were growing up.

Notes on Cooking Methods

- Fill in the chart corresponding to the cooking method that you researched.
- Describe the method and list the advantages of that method, followed by the disadvantages.

Cooking Method 1: _____

Description	Advantages	Disadvantages

Cooking Method 2: _____

Description	Advantages	Disadvantages

Cooking Method 3: _____

Description	Advantages	Disadvantages

Permission is granted to reproduce this page for purchaser's class only. Copyright © Trifolium Books Inc.

Radiation, Conduction or Convection?

- For each sketch, which word describes how the heat is being transferred?
- Be prepared to explain your answer in each case.

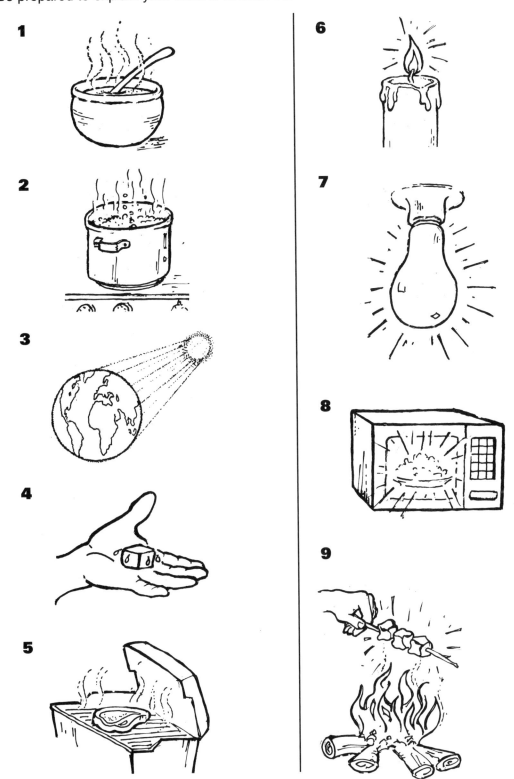

Chapter 3

Developing Language for Inquiry and Communication

Explaining Vocabulary

Demon words for all science students: "average, concept, contract, device, essential, factor, independent, minimum, partial, rate, source, theory "

(Henderson and Wellington 1998, Prophet and Towse 1999)

An understanding of vocabulary seems to be the most critical aspect of comprehension for English language learners. Recent research indicates that a reader requires an understanding of 95% of the words in a piece of text before the rest of the meaning may be inferred (Laufer 1992). The specialized vocabulary of science presents three obstacles: technical vocabulary, the use of words not normally encountered in conversation and learning the correct scientific meaning of words used differently in everyday speech.

Current studies on effective teaching practice de-emphasize the use of scientific terms. Studies have shown that students who can use technical vocabulary often do not understand the concepts they describe any better than their peers who refer to processes in simpler terms. Technical terms such as "buoyancy," " density" and "particle" may be no more difficult for English language learners than native English speakers as these are the terms and concepts that you will be teaching explicitly.

The stumbling blocks are more likely to be the formal vocabulary that is not part of youngsters' everyday conversation, words such as: determine, submerge, immerse, extremely, and relatively. The meanings of such terms also pose significant problems for many other children in the classroom (Henderson and Wellington 1998).

The multiple meanings of words in English create problems. English language learners know what a table is but may not understand "a table of contents" or the "water table." English speakers have probably encountered a word like pressure in the context of "I can't concentrate when I'm under so much pressure." They have to add the technical meanings for water and air pressure to their repertoire of meanings. English language learners have to learn the common meanings as well.

Dictionaries are essential for English language learners, but they are often inadequate. The tiny bilingual dictionaries students carry with them have a limited range of vocabulary; all languages do not have exact equivalents for words such as "nevertheless" and "consequently." The scientific meaning of a word like "mass" is unlikely to be listed first. Students often miss key parts of lessons as they search through their pocket dictionaries

> **Electronic Dictionaries:** Teachers and students disagree on the value of the expensive hand-held electronic dictionaries in language learning. Teachers deplore their use; the kids love them. A study done with Chinese students in three Vancouver schools (Tang 1997) found that electronic dictionaries are of very limited use to younger students. In order to find the English equivalent of a Chinese word, the student must know the number of strokes in Chinese, or be adept at romanizing Cantonese or Mandarin phonetically. The grade 8 students in the study had no dictionary skills. Even when they found the word, they could not understand the Chinese characters, and idioms were a wild goose chase. The dictionaries were even less useful when the students tried to find equivalents for their own writing, often producing hilarious results. In addition, many schools have prohibited the use of electronic dictionaries during examinations because some of them are programmable.

or type words into an electronic dictionary. Ironically, they may discover that the equivalent they find in their first language may be a word they are totally unfamiliar with as well.

A jigsaw crossword puzzle is a good way to review vocabulary and to convince students of the value of cooperative learning. This idea was developed by Elizabeth Coelho (Kessler 1992).

Activity

The Vocabulary of Fluids

Materials
Handout 3.1: Word Puzzle on Fluids
Handout 3.2: Clues for Word Puzzle on Fluids

- Divide the class into groups of 4. Distribute *Handout 3.1: Word Puzzle on Fluids* to each student. Then distribute a different set of clues, *Handout 3.2: Clues for Word Puzzle on Fluids*, to each student in the groups of 4.

- Instruct students to complete the word puzzle individually. When they express frustration, ask why. Agree that they don't have enough information but that as a group perhaps they do. Instruct them to tell each other their clues. Discourage them from exchanging papers or showing each other their information.

- Discuss the experience with the class. Most students enjoy the feeling of full participation for everyone in the class.

- Expand this activity by using student-prepared word puzzles for reviewing science vocabulary regularly.

Other Ways to Explain Vocabulary

People communicate frequently, if not always well. Hundreds of different languages have evolved to fit the needs of the people who use them. Because languages vary widely in sound, structure, and vocabulary and because language is so culturally bound, it is not always easy to translate from one to another with precision.

(American Association for the Advancement of Science 1993)

☞ Use examples from many cultures to relate a scientific word with an everyday item. When explaining viscosity, raid your kitchen for hoisin, soy, and fish sauces, salsas, as well as maple syrup, corn syrup, molasses, and ketchup.

☞ Encourage English language learners to use their Personal Dictionaries *(Handout 2.5, page 17)*. Maintaining an on-going bilingual word list is an essential method for learning vocabulary in another language.

☞ Create a science word bank, key word lists, and/or a classroom spellchecker. Display vocabulary on your classroom walls. Include common names of apparatus, important labelling words, words for important concepts and processes, and common units of measurement.

☞ Teach your students vocabulary in an innovative way using vocabulary tickets (Bassano and Christison 1992). When students are reading unfamiliar text, distribute scrap paper cut into ticket-sized pieces. Have students note words that are unfamiliar and deposit them in a container. Draw out words and explain them for the whole class. Students are often surprised at how many others are confused by vocabulary and appreciate this anonymous way of asking questions.

☞ Lead students through a session of vocabulary tic-tac-toe (X's and O's, crosses and noughts). Begin by matching nine vocabulary words from the class word bank to meanings. As an illustration, we have chosen non-technical academic language. Try a game with these words first.

Essential — necessary
Immerse — place completely underwater
Devise — create
Determine — decide
Extremely — very, very
Minimum — the smallest amount possible
Maximum — the largest amount possible
Factor — one of the things leading to a result
Significant — important

Community Connection

Quizzing students on bilingual word lists can be a family activity.

Word Puzzle on Fluids

- Use the clues on *Handout 3.2* for this crossword.
- Fill in the crossword using the clues assigned by your teacher.
- What does the mystery word mean?

Clues for Word Puzzle on Fluids

- Your teacher will assign you one of the 4 sets of clues.
- Read each clue and figure out the answer by yourself.
- Once you have solved all the clues, write down the meaning of the mystery word.

Student A

1. The amount of space a substance occupies
2. Its rate of flow can be measured
3. Type of mechanism used in car brakes
4. Liquids become this when cooled
5. Measures the density of liquids
6. Begins with "m"
7. Developed a principle about buoyancy
8. A measure of the average energy of the particles moving in a substance
9. Equals mass divided by volume

The mystery word is _____. It means _____

Student B

1. The unknown in *density = mass* divided by _____
2. It can be thick or thin
3. Confined pressurized systems that use moving liquids
4. Describes some liquids
5. Floats at different heights
6. This property depends on the object, not what it is made of or where it is located
7. Told the king his crown was not pure gold
8. Can be measured using a thermometer
9. A physical property of a substance

The mystery word is _____. It means _____

Fluids — Chapter Three — Handout 3.2b

Student C

1. Starts with "v"
2. Includes gases
3. Type of mechanism used to lift heavy objects
4. Ends in "k"
5. A device marked with a measurement scale
6. The property of an object that resists a change in its motion
7. Famous for yelling "Eureka!"
8. When it changes, the rate of flow of a fluid changes
9. Explains why oil floats on water

The mystery word is _____. It means _____

Student D

1. Pressure can change this property of a fluid
2. Includes liquids
3. Starts with "h"
4. These types of fluids flow slowly
5. Has a mass at the bottom
6. Can be measured in grams
7. Greek mathematician and scientist
8. Starts with "t"
9. Measured with a hydrometer for liquids

The mystery word is _____. It means _____

Words that Matter Tic-Tac-Toe

- Review 12 words and meanings in your group.
- Copy the meanings of nine of these words in the spaces provided. Use any order you wish.
- Have one of the group members put the words you are reviewing on slips of paper in a box.
- Decide whether finishing the game involves making straight lines or filling the entire grid.
- Choose a group member to read out the words.
- When you hear a word, put a marker on the meaning.

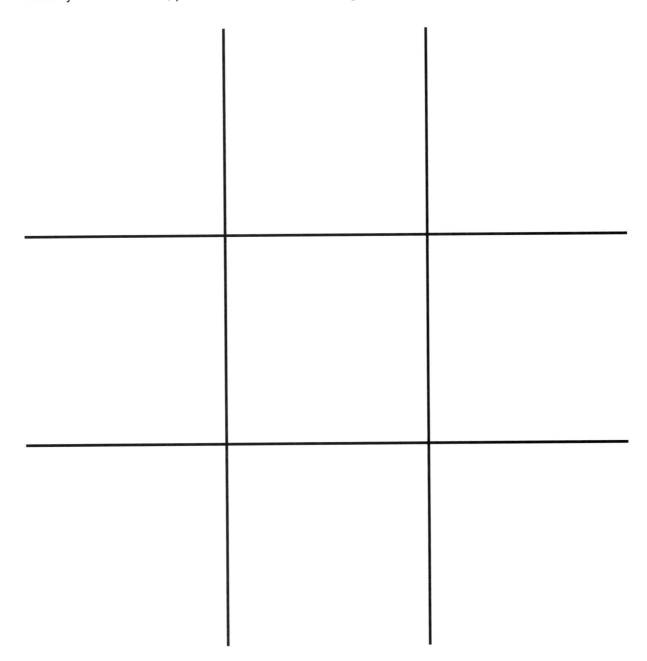

Practising Asking Questions

The formulation of a problem is often more essential than its solution which may be merely a matter of mathematical or experimental skill.... To raise new questions, new possibilities, to regard old problems from a new angle, requires creative imagination and marks real advances

(Albert Einstein) – quoted in Costa, Kallick 2000

When learning a new language, students, especially those raised in traditional school systems, yearn to get the right answer. The most inquisitive may be hesitant to risk asking questions. It is often difficult for them to comprehend that the essence of science is posing questions and framing hypotheses to be tested.

The syntax of questions in English can be daunting. First the speaker must select one of the hard-to-pronounce, almost identical interrogative pronouns (e.g., which, what, who) or an interrogative adverb (e.g., how, where, when, why). Then comes remembering to reverse the order of the subject and the verb, a process that toddlers find difficult and instead ask "Where Daddy is?" with a quizzical expression. A further complication is deciding if the verb should be contracted as in "What's the pressure?" Or if a "yes" or "no" answer is expected, the question should be phrased without an interrogative pronoun as in, "Is it hot?" Until these question forms become automatic, English language learners have justification for shelving their curiosity. Further difficulties arise if the syntax of the question requires the passive voice.

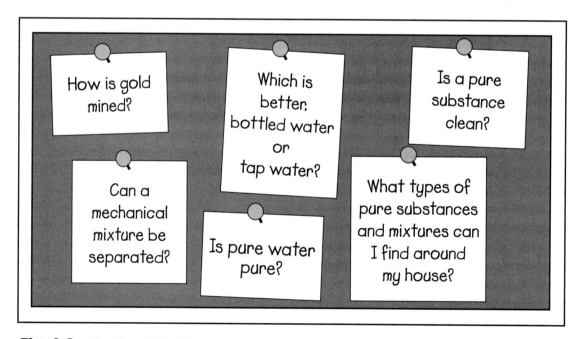

Fig. 2.8 Wonder Wall. *The word "wonder" has multiple meanings, for example: I* **wonder** *what life is like on Mars. Watching a sunset always fills me with* **wonder**.

One way to meet the inquiry demands of the science curriculum and ease English language learners into the complex language required is to use a board or cork board as a centre for recording questions that arise. Some teachers call this a Wonder Wall.

This is an ideal strategy for tracking those unexpected questions that may not be totally relevant when they come up but need to be addressed later in the unit. Written questions posted on the Wonder Wall serve as models of interrogative forms for English language learners.

Questions about Pure Substances and Mixtures

Materials
questions on Wonder Wall
coloured dots

- After you have introduced the unit topic "Pure Substances and Mixtures," encourage students to formulate questions to be posted on a Wonder Wall. See Fig. 3.1 for some suggestions.

- As the class learns the answers to some of the preliminary questions, remove these but continue to add new questions throughout the unit.

- To review this unit, try a classifying activity. Give students coloured dots and ask them to identify which questions may be answered by consulting a textbook, which would be suitable for an experiment and which may have no answers.

Other Ways to Practise Asking Questions

Inquirers recognize discrepancies and phenomena in their environment, and they probe into their causes: Why do cats purr? How high can birds fly?

(Costa and Kallick 2000)

☞ Encourage speculative thought and practise the forms used in cause-and-effect statements and hypotheses statements with a game such as "What would happen if...." *Handout 3.4: What if...* lists some speculative questions on lab safety.

☞ The Science Centre of Singapore has a ScienceNet section on its web site **www.science.edu.sg/ssc** where students can review interesting questions such as: What is the process of making a century egg? What is the chemical absorbent used in disposable diapers, and how does it work?

Have students draft class review questions prior to unit tests.

The Rules for "if" Clauses: *Use the past tense of the verb in an "if" clause and the conditional in the main clause. If you gave me a million dollars, I would take the whole class to Disneyworld.*

What if ...

- Choose a partner for this activity.
- Cut the boxes apart and put them face down between both partners.
- Partner 1 picks up the first card and reads it to Partner 2. Partner 2 answers the question by starting with "If _____, then _____."
- Next, Partner 2 reads the next card to Partner 1 and Partner 1 answers it.
- Repeat the previous two steps until all the questions have been read out and answered.

What would you do if you broke a piece of lab equipment	What would you do if your partner's hair caught on fire?	What would you do if you splashed chemicals in your eye?
What would happen if you forgot to put on your safety goggles?	What could happen if you did not tie your long hair back while using an open flame?	What would you do if you spilled some chemicals?
What should you do with your chemicals after you finish your experiment?	What could happen if you fooled around during an experiment?	What could happen if you didn't listen to your teacher's instructions before an experiment?

Working with Word Problems

Word problems ... are often difficult for newcomers, because they have to 'decode' unfamiliar content and terminology before they can begin to apply mathematical skills to the problem.

(Coelho 1998)

The reading techniques required to understand word problems are very different from the skimming and scanning approaches required in most reading. A problem provides very little context; the narrative is an embellishment more than an essential; the natural redundancy of spoken language is missing. Students must read every single word.

When assigning a set of word problems, consider the background knowledge students must have to be able to solve them. Contexts, especially those involving sports, are confusing. Imagine trying to solve a problem like the one in the margin. Would it be easy or difficult for you? Why? Although the arithmetic in the question is simple, it may be impossible for some people to solve the problem if they are not familiar with golfing terms. Sometimes the words which indicate the mathematical step (e.g., increase, the sum, twice as) may be unfamiliar to English language learners. Visit **www.hbschool.com/glossary/math** for helpful definitions.

It is important to keep in mind that word problems are difficult for all students. For this reason, avoid ambiguities or unnecessary information in the problems. Keep the wording concise and as direct as possible. Translating words into mathematical statements is very difficult for most students, so avoid sophisticated or lengthy word problems. Lastly, make it very clear what items you are looking for when awarding marks; students often think that just having the correct numerical answer is sufficient to get full marks. They do not understand that methodology, which allows them to tackle more difficult problems, is far more important than the final number itself.

One of the ways to help students with problem-solving is to model the reasoning process using "a think aloud" approach. Try this technique with the mechanical efficiency problem in the activity below.

It was such a nice day that Chris decided to spend the afternoon on the links. After half a round on a par three course, she had three bogies, one double bogie, a birdie and an eagle on her score card. The other holes were on par. What was her score?

Activity

A "Think Aloud" Mechanical Efficiency Problem

Materials
Handout 3.5: Steps in Solving Word Problems (to help students understand the steps)

- Write a problem like this one on the board or an overhead transparency:

  When re-painting a spot on the back of a gazebo a few days after he first painted it, Jamal had to re-open the can of paint he had originally used. Since he had banged the can closed with a hammer and some of the paint had dried around the rim, he estimated the amount of force needed to lift the lid off the can to be 100 N. He estimated that the effort needed to open the can would be 4 N if he used a screwdriver as a lever. What would be the mechanical advantage of using the screwdriver?

- Talk your way through the problem like this:

  My goodness, I don't even know what a gazebo is. Maybe it doesn't matter. There is a lot of information about painting and the can but the question seems to be about mechanical advantage. Mechanical advantage is the ratio of the load force divided by the effort force. In this question, the load force is 100 N because that is the amount of force needed to lift the lid from the can. However, Jamal estimates that he only needs to apply a force of 4 N if he uses the screwdriver as a lever. The mechanical advantage is 100 N divided by 4 N which is equal to 25.

Other Ways to Work with Word Problems

Quite often the simple act of talking through the problem not only yields a correct solution but, in the long run, gives students critical practice in learning and thinking in the academic language of English.

(Chamot and O'Malley 1994)

☞ Use a mnemonic device such as "GRASS" to help students remember the steps for problem-solving to be used with all grades in the school. Post these steps in the classroom. *Handout 3.5: Steps in Solving Word Problems* is an example. Have students who are proficient in English work through the steps with English language learners until they can verbalize the steps in English. GRASS could become a standard procedure for problem-solving used throughout a school.

☞ For problems written on the board, have students take the time to copy each problem carefully. Illustrate how to stroke out information they think is unnecessary. Always encourage them to make sketches to visualize a problem especially if it is a physics problem, such as the direction of forces in pulleys. This can help you understand their thinking process as they solve the problem. Have them write on their sketch all the information that they know in mathematical terms, whether it is explicitly stated in the problem or implied.

☞ Form groups so that the mathematically skilled and English proficient students are evenly distributed. In each group, include a student who is fairly knowledgeable about common objects or activities in your area, such as the terminology associated with a particular sport or parts of a house. Display the problem on an overhead projector, or give each student in a group only one piece of information about the problem. If groups finish quickly or become frustrated, encourage a member of each group to offer help to other students or to ask you for assistance.

☞ Raise awareness of problem-solving strategies by having students fill out a private survey such as the one in *Handout 3.6: A Private Survey*. Monitor student progress by checking to see that they are implementing the ideas they wrote down on their surveys.

☞ Help students understand how problems are constructed by having them write their own from data provided.

GRASS (Steps in Solving Word Problems)

- Follow these steps when you are solving word problems.
- Make sure that you consider the questions written below each step.

Given

- Can I draw a sketch for this problem?
- What are the important facts?
- How do I write these facts mathematically?

Required

- What question am I answering?
- What quantity am I asked to calculate?

Assemble

- Is there a formula that I know that relates the facts in the Given with the quantity in Required?
- What is the formula?
- Do I need to break down the problem into smaller problems first and calculate something else first?

Substitute

- What units do I need to use in the formulas?
- Do I have to convert any of the facts in the Given into the correct units?
- What numbers do I put in the formulas?

Solve

- Have I done all the calculations and checked my work?
- Is the answer realistic for the situation described in the problem?
- Do I have to round off my final answer?
- How do I know how many digits should be in my final answer?
- Have I included a concluding statement?

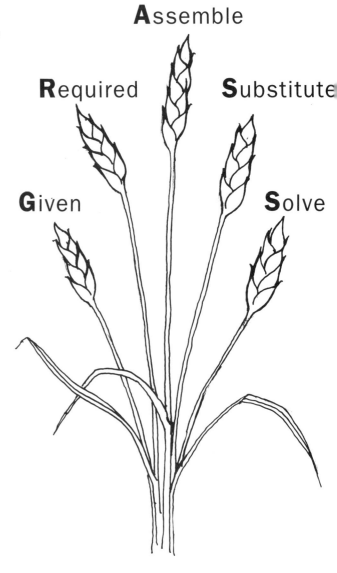

A Private Survey

This quiz is only for you. Don't show it to your teacher or your friends. There are no right or wrong answers.

- Think about the times when you had to do a problem in science class.
- Think about each statement and colour the appropriate squares, starting from 1, to show your answer.
- Think about which of these statements are good ideas and which are not helpful.
- Write a paragraph in your science journal explaining how you will use one of the strategies you don't use now.

	Never				Sometimes					Most often
	1	2	3	4	5	6	7	8	9	10
I always read problems carefully.										
I copy out the problem when I'm trying to understand it.										
I draw a little diagram or make a chart to help solve problems.										
When I don't understand some words, I give up and stop working on the problem.										
I follow a plan to solve problems.										
I talk the problem over with a friend.										
I copy my friend's homework.										
I make mistakes in calculations.										
I check my answers carefully.										

Permission is granted to reproduce this page for purchaser's class only. Copyright © Trifolium Books Inc.

Guiding Written Responses

When the purpose of science education is primarily as training for future scientists, learning to present evidence in the style of a scientific paper is valuable. However, for the majority of pupils, the scientific language may hinder their understanding of science.

(Henderson and Wellington 1998)

Young scientists need to write in order to learn. We must be careful not to exempt English language learners from using writing as a way of understanding information. All writers require guidance as they progress from writing simple narratives to recording and describing scientific information. English language learners may be familiar with rhetorical styles that differ from English, such as keeping the most important points for the very end. It helps to supply sentence and paragraph starters, sub-titles and concluding sentences.

In science we tend to focus on formal written reports of practical work. The use of the passive voice in scientific observation is very confusing, especially for speakers of languages that have no equivalent. The current trend of scientific writing is to be more personal and to use expressions such as "We noted that" instead of "It was observed that." All students understand this way of writing more easily.

Narrative and dialogue also help students learn. We discussed one example of using narrative in Chapter 2 using the topic of cells and *Handout 2.3: A Day in the Life of a Unicellular Organism* on page 15.

The activity "A Script on Electricity" involves writing a script. Choose a video or film intended for viewers without much knowledge of electricity. Your goal is to use the film frames themselves as prompts by working with the film excluding the soundtrack and having the students produce a script.

Activity

A Script on Electricity

Materials
10 min video-clip
a K.W.L.H. chart (see *Handout 2.2* on page 11)

- Show students a brief (less than 10 minutes) video-clip twice without the sound.
- After the first viewing, discuss what they thought the video was about. Explain that they will see the film again and that this time they should look for scientific ideas.

- After the second viewing, help students organize the information they have identified using a K.W.L.H. chart on the board.

- As a class, use the chart to write a common script for the video.

- After you have taught several lessons on electricity using demonstrations, experiments, textbook readings, concept and word lists, etc., show the video again. Pause for discussion after each section.

- Have each student write an individual draft of the video-script.

- Show the video-clip again, this time with sound. Give students an opportunity to confer with partners. Have partners indicate what essential information may have been omitted.

- Have the students complete the K.W.L.H. chart and write a summary in their science journals.

Other Ways to Guide Written Responses

Writing to learn activities are not the place to work on spelling, grammar or sentence structure.

(Susan Krischner quoted in Rosenthal 1996)

☞ Use point form graphic organizers as a scaffolding for writing. For the topic of electricity, you might use a flow chart to relate different types of electrical circuits, a Venn diagram for comparing sources of power, a T-chart with main ideas on one side and supporting details on the other for paragraphs on conserving electricity, and a pictorial concept map for a summary of what students have learned in the unit.

☞ Have students write down new concepts they have learned, questions that they have, and other interesting science-related comments in their science journals. These comments and questions can serve as an idea bank for writing topics for the whole class.

☞ Help students learn to rewrite a question as a statement that forms the topic sentence of a written answer, for example:

What is static electricity? Static electricity is … ;
Where does your electricity come from? Our electricity comes from … ;
What are three key discoveries in the history of electricity? The first discovery in the history of electricity was …. Another was …. A third was ….

Making Textbooks Accessible

Textbooks (if schools can afford such a luxury) are often used to provide homework, to guide a practical [experiment], to keep pupils busy if they finish too soon, or, at worst, to prop up a piece of equipment

(Henderson and Wellington 1998).

The listening and doing aspect of good science classes makes science an excellent language learning opportunity for English language learners. But if the concepts and language explained are to be consolidated, students must also absorb them from print.

The newest textbooks are seductive. They are beautifully illustrated with a wealth of supplementary activities laid out in an attractive magazine format. Readability formulas have been applied. But English language learners find it hard to figure out what part of the page they are meant to read and which diagrams and photographs are important. It doesn't matter how many words there are per sentence if they don't know many of them. In short, "the features that sell the book may have little to do with how helpful it actually will be to the students who are supposed to read and learn from it (Rosenthal 1996)."

Textbook reading should be teacher directed and is most valuable when it is introduced *after* students have an initial understanding of the content, never before. When you have introduced a topic, explored students' prior knowledge, explained and used much of the vocabulary, and perhaps done a demonstration or an introductory experiment, organize a directed reading of the text. The familiarity of content helps students gain confidence in the subject and in their own abilities to read and understand a textbook.

Activity

Analysis of a Textbook Reading on Optics

Materials
an overhead transparency of 1 textbook page
Handout 3.7: Steps for Reading
another textbook page for students to read

- Make an overhead transparency of one page of the textbook you are using to model the steps students should follow when reading.
- Cover the passage with a piece of paper showing only the title.

Read the title to the class and ask what they predict the reading will be about.

- Then have them consider a key illustration and the caption beneath it. Ask the class how this connects to the title.

- Read the passage aloud to the class. Turn off the overhead and elicit what information they found to be unusual, interesting or important.

- Distribute *Handout 3.7: Steps for Reading* and assign a suitable reading for students to work with in pairs.

Other Ways to Make Textbooks Accessible

The information is dense and often presented with little repetition or paraphrasing. The emotion, drama and chronological unfolding of events that characterize many other forms of writing and other subject areas are absent from the science text, often making the reading tedious and slow-going.

(Rosenthal 1996)

☞ Read aloud from the science text as often as you can find time: to the class, to small groups, and to individuals. The modelling of pronunciation and phrasing assists students in comprehending material that is too difficult for independent reading. English language learners may have understood your explanation of phosphorescence and bioluminescence but flounder when they have to decipher these words in print.

☞ Go over the textbook pages before you teach the class to see what material is relevant to the lesson. Often textbooks contain far more information than can possibly be taught in a school year. Be very

Avoid requiring students to read aloud. Besides being embarrassing, reading aloud can shift the mind away from content to concern with pronunciation.

Pages to Study:

Re-read pages 103 and 104,
yellow sidebar on page 105, illustration at top of page 106.
OMIT pages 107-10.
Study white parts of pages 111-14.
Learn definitions for these words in bold type: **lens, refract, convex lens, concave lens, principal axis, focal length.**

selective about what you want students to read, and always ask yourself: "Is this material important for my students now?" "Is it presented clearly and concisely with appropriate visuals?"

☞ Be clear about what parts of the textbook they are responsible for studying. Always write down the page numbers of textbook material that students are responsible for on the board as shown on the opposite page. Most students understand things better when they see things in writing rather than listen to a list of page numbers. Make sure you give your students time to copy down the page numbers in their notebooks.

☞ Use the teacher resources that come with class sets of texts. Key illustrations are often available as overhead transparencies which are excellent for focusing pre-reading discussions. Do not be afraid to use other teacher resources that your school may have from previous years; you may find very useful student readings and transparencies not included in your most recent resource.

☞ Set definite tasks like these:

(a) Make up cards of true and false statements for a few textbook pages. Have students predict which are true and false *before* they read by sorting the cards into true and false piles. Then have them read with a partner to confirm and revise their decisions.

(b) Photocopy and cut up a paragraph from the text. Give a different sentence to each student in a group and have them figure out the sequence. Then have them check their sequence with the text.

(c) Provide students with graphic organizers to fill in such as Venn diagrams for comparisons (see *Handout 2.10* on page 29 as an example), flow charts to explain a sequence or cause-and-effect relationships (see *Handouts 2.8* and *2.11* on pages 27 and 30 as examples), and T-charts to identify main ideas and supporting details.

(d) Have pairs of students make up questions or suggest suitable tables or charts to summarize information.

Community Connection

Community groups may be able to provide copies of science textbooks in first language. These will help students build the cognitive framework for learning similar material in English.

Steps for Reading

1 Read the title.

2 Look at the pictures or diagrams and read the words under the pictures (called the captions). What do you think the reading will be about?

3 Read through the passage silently and quickly. Do not slow down if you don't know a few words.

4 Turn over the book. Talk with a partner about the most interesting, the strangest or the most important thing you found out.

5 Reread the passage to check what you know. Read the first sentence and last sentence of each paragraph very carefully. You can often understand the meaning of a paragraph by reading only the topic and concluding sentences.

6 If there are questions with the passage, read them.

7 Decide whether you would like to have the teacher read the passage aloud to you.

8 Read it once more and answer the questions or make up questions for your classmates to answer.

# Forces

Chapter Two

Chapter 4

Making Connections to the World Outside the Classroom

Implementing Inclusive Curriculum

All human cultures have included study of nature—the movement of heavenly bodies, the behavior of animals, the properties of materials, the medicinal properties of plants.

(American Association for the Advancement of Science 1989)

Some of the most awe-inspiring photographs of the twentieth century are those of Earth from space. That image of a fragile blue ball sustaining billions of people has inspired an environmental effort to save the planet. Children schooled in the twenty-first century need a global perspective. English language learners connect the science they learn in the classroom, not only to their new environment but to the environment they have come from. If you can encourage these links and tap into them, such first hand experiences will enrich the curriculum for the whole class.

As students begin their study of space, recognize that from earliest times humans have tried to understand what they saw in the night sky. Cultures throughout the world used different names to describe the constellations. The group of stars we call the Big Dipper in North America is known to Native Americans as the Great Bear, to the British as the Plough, and in the past to the Chinese as the Celestial Bureaucrat. Other cultures have stories about the stars and the creation of the universe. Be aware that a common experience like viewing the night sky may be radically different from other points on Earth or may have been totally inaccessible to students who were growing up in areas where curfews were imposed.

Looking at science from a historical perspective helps students understand that scientific knowledge is the result of studies from all over the world being carried out over a long period of time. Middle school is the first time such an approach is possible; detailed study is not recommended because young adolescents have not studied enough history to be able to fully understand contexts. It is important to stress the value of contributions from people all over the world and to emphasize that theories that are later proven invalid are essential to the progress of scientific thought.

One way to ensure that English language learners participate is to provide a simple scaffolding for a discussion on the nature of evidence.

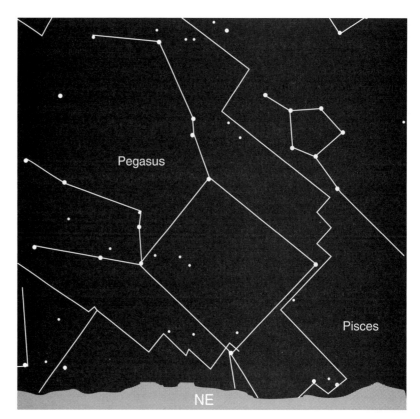

Fig. 4.1 *View of night sky from Buenos Aires, Argentina.*

Activity

Analysis of Scientific Reasoning

Materials
Handout 4.1: What Proof Would You Accept?

- Divide the class into groups of 4 and give each group a copy of *Handout 4.1: What Proof Would You Accept?*

- Explain to the students that they should not discuss whether they believe the statement is true or false, but they should decide what evidence would convince them that the statement is true. If the students cannot agree on acceptable proof, they should write down their differences of opinion.

- When all the statements have been discussed, the **groups** report back to the class.

Other Ways to Make Science Curriculum Inclusive

Scientific work involves many individuals doing many different kinds of work and goes on in some degree in all nations of the world. Men and women of all ethnic and national backgrounds participate in science and its applications.

(American Association for the Advancement of Science 1989)

☞ Encourage students to share stories about the constellations from different parts of the world. Biographies of astronomers such as Galileo Galilei, Maria Mitchell, Helen Hogg, and Carl Sagan or of astronauts from around the world such as Yuri Gagarin, Ellen Ochoa, Julie La Payette, and Marc Garneau are often available in versions that are accessible to readers who are new to English.

☞ A way to reinforce the idea that throughout history men and women from many parts of the world have contributed to current scientific thinking is through constructing time-lines. A historical study of attempts to design spacecraft is more age appropriate than theories about the nature of the universe.

☞ Emphasize the importance of teamwork in scientific progress. A great example that illustrates the progress that various individuals made in science deals with determining the shape of the Earth. Eratosthenes of Cyrene (276–196 BCE) was a Greek mathematician, geographer, astronomer, and philosopher who was appointed head of the Alexandrian library in Egypt. Using an observation that someone else before him had made about the length of shadows in Syene, Egypt on the summer solstice, Eratosthenes developed an experiment which eventually showed that the Earth was round, not flat as was previously believed. In fact, he calculated the circumference of the Earth with an error of only a few percent, using mathematical tools that had been developed by various people before and during his time. This remarkable achievement was almost forgotten until in 1522 del Cano, captain of one of Ferdinand Magellan's ships, became the first person to circumnavigate the globe.

☞ Encourage students to share their hunches and questions. The Wonder Wall described in Chapter 3 on page 44 is one way to do this. Make sure there is an appreciation of the intuitive breakthroughs in science as well as an understanding of the scientific method.

☞ One of NASA's web sites, **starchild.gsfc.nasa.gov**, offers information about the sun, the moon, the solar system, etc. at two different reading levels with an excellent clickable glossary. This information is labelled printable and is also available to teachers as a free CD-ROM.

"I felt as if I came from a backward place."

What Proof Would You Accept?

- For each statement in the first column, discuss with your group what information you would need in order to believe that statement to be true.
- Write down your responses beside each statement.

Statement	Proof We Would Accept
The Earth is round.	
There is life on Mars.	
Aliens in space ships visit Earth regularly.	
Other planets have moons of their own.	
The moon is full on the twentieth day of every month.	
Summers have been getting hotter since 1990.	
Looking directly at the sun during an eclipse damages your eyes.	
Before clocks were invented, nobody knew what time it was.	
High tide occurs at the same time every day.	
There are mountains on the moon	

Using Authentic Reading and Viewing Materials

Once people gain a good sense of how science operates—along with a basic inventory of key science concepts as a basis for learning more later—they can follow the science adventure story as it plays out during their lifetimes.

(American Association for the Advancement of Science 1993)

Authentic material is not written for the classroom but for the general public, and may consist of brochures, newspapers or magazines. The media is a powerful tool for raising awareness in the general public about environmental problems which affect water systems. Classroom texts, printed a year or so after they are written, are only one source of information. It is important to include English language learners in investigations of current events rather than reserve these explorations for enrichment. Following a news story such as a flood or drought can be a powerful way of explaining the water cycle. Usually such stories are covered in the first language press and are updated on newspaper internet sites around the world.

Television news clips announcing discoveries and exploring issues (e.g., water conservation during droughts, contaminated water) or describing disasters (e.g., floods and oil spills affecting water systems) are valuable. The CBC News in Review is a monthly compendium of short news clips of current events related to the school curriculum, such as the ice storm in eastern Canada in 1999. In the United States, CNN has a web page with transcripts of current programming. Because commentary is conversational, a transcript may be accessible to less proficient readers. Material is available in a variety of languages including Spanish and Korean.

Collections of non-fiction books, magazines, daily news clippings and children's encyclopedias can also be valuable sources of information for students who are reading below grade level.

Global topics that lend themselves to research are where community water supplies come from and how people conserve water.

Activity

An Interview on Water Supplies

Materials
Handout 4.2: An Interview with an Adult on Water Supplies
materials to make posters

- Assign students the task of interviewing an adult in their home or community about water supplies in various parts of the world. Have them use *Handout 4.2: An Interview with an Adult on Water Supplies* as a guide for their questions.

- Have students communicate the information they have gathered by making posters on water conservation around the world.

- Each poster should have three interesting facts, an illustration and a title.

Other Ways to Use Authentic Reading and Viewing Materials

Simplifying materials denies ESOL students access to the very language they need, and limits their opportunities for new learning. Giving ESOL students more than one word, more than one reading, or more than one avenue for understanding a concept can be more useful than watering down content.

(Teemant, Bernhardt and Rodriguez-Munoz 1996)

"The world is our classroom."

☞ Use pamphlets issued for community education by local water authorities. They are often written in clear language that members of the public can understand. In areas with large immigrant populations, such material is often available in a variety of languages.

☞ Have a variety of popular non-fiction library books available during silent reading sessions. English language learners can build scientific concepts and language from their independent reading of illustrated factual books intended for younger readers.

☞ Make displays from newspapers. Students can provide local first language newspapers. When the reading level seems difficult, try matching pictures with captions and headlines or have students write their own headlines.

☞ Have a committee of students clip feature science columns from newspapers and magazines. Other committees could regularly scan TV guides for programs such as those on the Discovery Channel, TV Ontario, and PBS (Public Broadcasting System) that could be assigned for homework.

☞ English language learners should be encouraged to use closed captions, the English subtitles of the dialogue. Since 1993, all televisions with screens larger than 33 cm (13 inches) have been manufactured with built-in decoders.

Families can help in assembling newspapers and pamphlets in first language

An Interview with an Adult on Water Supplies

- Choose an adult to interview on the topic of water supplies from their country of origin.
- Review the list of starter questions below.
- Prepare a list of specific questions you have of your own in addition to the starter questions.
- Use the starter questions and your own specific ones for your interview.
- Make notes during the interview.

Starter Questions

1. Where did you grow up and what cities, towns, or villages have you lived in?
2. Where did your water supply come from?
3. How difficult or easy was it to get your water?
4. Was the drinking water safe? How did your family make sure that was so?
5. Did you have to be careful about conserving water? How did you do that?
6. How did the water come to your home?
7. Was it the custom to go camping or travelling? What extra care was needed with drinking water?
8. Did people swim? Was this safe? Why or why not?
9. What problems did people have with the water supply?
10. How did weather affect the water supply?
11. How was waste water dealt with?
12. Were there any human practices that affected the safety or quantity of water? If so, what were they?
13. Did the local government have any responsibility in regulating water supplies? If so, what did they do?

Permission is granted to reproduce this page for purchaser's class only. Copyright © Trifolium Books Inc.

Constructing Working Models

The usefulness of conceptual models depends on the ability of people to imagine that something they do not understand is in some way like something that they do understand.

(American Association for the Advancement of Science 1993)

Creating a model is an excellent way for English language learners to demonstrate their understanding of concepts even when their linguistic ability may be quite limited. Moreover, they may be able to create prototypes of toys or vehicles. The whole class needs to understand that models are alternative representations of reality. They are used because the object is too large, as is the solar system; or too small, as is an atom; too inaccessible, as is a skeleton; or too dangerous, as is an electrical circuit.

Middle school students can use models during their study of motion. They can adapt physical models to observe the effects of using different materials, adding more wheels, or reducing the mass. Constructing models provides English language learners with a non-verbal way to demonstrate their understanding of motion.

Activity

Design of a Mechanical Model

Materials
several mechanical models as examples (e.g., a handbeater, a levered corkscrew, clothesline with a pulley)
wheels, gears, pulleys, popsicle sticks, etc.
Handout 4.3: Planning Sheet for Mechanical Models

- Begin by bringing a variety of mechanical models into class so that the students can understand what their assignment is.

- Identify some of the materials that have been used in making these models. Consider supplying some materials such as wheels, gears, pulleys, popsicle sticks, etc. as well as encouraging students to bring things from home if possible.

- Identify a theme for the models to be built. For example, you may focus on making everyday life simpler around the house. You may set-up a model of a clothesline using pulleys, string and two small poles. Ask the students why someone would want a clothesline with pulleys rather than a stationary model. What are the advantages and disadvantages?

- Challenge the students to think of other ways to make things move around the house and make life more convenient. Have them plan using *Handout 4.3: Planning Sheet for Mechanical Models.*

- Have each student construct a simple model of their idea (either a real device like the clothesline or a device they invent) and demonstrate it to the rest of the class.

Other Ways to Construct Working Models

Dalton later made wooden models of atoms, and I saw his actual models in the Science Museum as a boy. These, crude and diagrammatic as they were, excited my imagination, helped give me a sense that atoms really existed. But not everyone felt this.... "Atoms," the eminent chemist H. E. Roscoe was to write, eighty years later "are round pieces of wood invented by Mr. Dalton."

(Sacks 2001)

☞ Ask students to design toys or other devices incorporating specific simple machines or mechanisms such as ramps, levers, gears, pulleys, etc. Have them consider the appropriateness of certain materials for the type of mechanical device they intend to use.

☞ Have students create models from specific periods in history or geographic locations. Students will likely require some information on what materials and technology was or is available during that time or place. Consider assigning a short research project before they undertake building these models.

☞ Extend the model building activity by having students create models of mechanical devices used in specific locations such as a farm or circus. Ask them to consider why these devices are used and what materials they are made from.

☞ Students can also build models that focus on hypothetical scenarios such as devices that may exist in the far future. Ask students to explain why their models would be useful and what materials may be required.

Planning Sheet for Mechanical Models

- Before building a model, complete steps 1 to 4.
- Build your model and complete steps 5 to 7.
- Could you improve your model? How would you do it?

1. What I want to build …

2. What I have to build with …

3. What I want it to look like …

4. How I want it to work …

5. What it actually looks like …

6. How it actually works …

7. Why it is useful …

Organizing Community Field Trips

Field trips can reinforce a particular unit of study, expose students to United States culture, motivate students, stimulate intellectual growth, and provide students with opportunities to receive meaningful input and to interact in real world situations.

(Scarcella and Oxford 1992)

Fieldwork is one of the areas in which the demands of studying science and learning a language converge. The concrete experiences students share build a knowledge and language base in a highly pleasurable way. This enjoyable aspect may mean that new students and their parents or guardians perceive field trips as recreational outings rather than an empirical method to learn curricular material. It is important to be explicit about what you expect the class to learn.

For field trips outside the school, be very clear about suitable clothing; students may be accustomed to wearing their best clothes on occasions when they are representing their school and find themselves shivering and wearing unsuitable shoes for a nature walk. Make plans for lunch both in terms of budget and possible dietary restrictions; always present choices for students if they do not bring their own lunches.

Often permission letters for school excursions must fulfill legal requirements, and these legalities make them incomprehensible to parents or guardians. An attached note explaining the purpose and practicalities of the trip with instructions to sign the legal letter helps. Always check to see what your school requires you to include in the letter to parents or guardians. Try to write letters explaining excursions simply (e.g., *Handout 4.4: Permission Letter for Community Field Trips*) and if possible have such letters translated into the languages of your students (e.g., *Handout 4.5: Permission Letter for Community Field Trips in Spanish, Handout 4.6 Permission Letter for Community Field Trips in Chinese*).

Overnight trips, when well managed, can be the highlight of the school year. However, parents or guardians may be reluctant to entrust their children to your care. A special meeting for parents and guardians with translators in which you show slides of other such excursions helps. Emphasize the educational importance of the trip and be careful to include detailed explanations of the level of supervision, the separation of male and female sleeping quarters, the provision of appropriate food as well as an emphasis on the educational importance of the trip. One school, because of space constraints, divided their week at an outdoor education center into girls' days and boys' days. The teachers were overwhelmed by the improvement in the behaviour of both groups and, incidentally, in the increase in numbers of English language learners who were allowed to join their classmates.

Planning field trips within the schoolyard or within walking distance of the school increases the amount of time your class can spend on an activity that is valuable for English language learners. The activity "Field Trip to Study Living Organisms" is borrowed from *Take an Ecowalk to Explore Science Concepts 1*.

Activity

Field Trip to Study Living Organisms

Materials

clipboards (or pieces of cardboard with elastic bands)
trowels, magnifying glasses
Handout 4.7: A Bug's Eye View

- Preparation: If possible, provide students with clipboards. Not only is recording data easier but young people who are clearly doing schoolwork are more warmly welcomed in the community. If clipboards are not available, pieces of cardboard with elastic bands can be substituted.

- Plan the field trip as a walk through an area that has a diversity of living organisms. If the area is close to your school, students can spend more time observing and participating in the field trip.

- Students who are fearful of spiders, snakes, and worms need to be reassured and taught that all living things are part of our environment and should not be harmed.

- Before the walk, discuss what is to be observed. For example, if you are going out to observe earthworms, discuss the role of earthworms as decomposers in a soil ecosystem. Also, introduce students to any new equipment (e.g., trowels, magnifying glasses) and to the recording sheets which will be used during the walk.

- During the walk, make sure students have equipment such as trowels to dig holes in order to find earthworms and a place to put specimens for observation.

- Use magnifying glasses to get a close-up look at specimens.

- Ask students to record their observations on *Handout 4.7: A Bug's Eye View*.

- Make sure students return the specimens unharmed to their environment.

Other Ways to Organize Community Field Trips

Teaching takes time, more time than is typically available at school. Designing activities that extend investigations beyond the classroom and into the community can enrich science learning while promoting access to English.

(Fathman, Quinn and Kessler 1992)

☞ Students can learn more about the diversity of living organisms if you plan field trips in various environments. Plan a walk through all parts of the school grounds to study the wide variety of flowers, plants and trees. Walk to a nearby ravine or forested area to study the variety of trees. Walk around the neighbourhood and collect information about the types of plants local residents use in their front gardens.

☞ Visit the local public library to find out what topics your students could research. Discuss the possibility of having resources from the children's collection tactfully displayed with age appropriate material. Sometimes libraries can arrange to provide library cards to new arrivals if alerted in advance. Public libraries also have workshops and other activities that can be combined with material your students are using in the classroom; discuss these opportunities with the librarian.

☞ Consider repeating an outdoor activity in another season. This allows students to see that learning is on-going and the information they learn is not static.

☞ All day visits to local zoos, science centres or natural science museums are musts when considering the diversity of living organisms. Many of these institutions have outreach programs with speakers who will bring specimens and slides to your classroom to prepare students for your trip.

"I remember the canoe trip. All those bugs and worrying about bears. I still have the pictures."

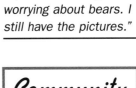

Enlist community volunteers as supervisors for field trips. Encourage students to return to field trip sites with their families

Permission Letter for Community Field Trips

Dear Parents and Guardians,

During the school year, the students in our class will be going out to do fieldwork in the schoolyard and the area near the school. We will study scientific topics and discover how science is important in the real world. We hope to do this work at least 10 times this year.

I will explain the science idea first, and follow up with a discussion and other activities in the classroom. Please discuss these activities with your child. We would also like to invite you to visit our classroom or join us on our excursions.

Thank you.

Please complete and return the form below with your child.

I give permission for my daughter/my son, _____, to leave school property for science excursions with the class. I understand that there are possible risks involved. I will remind my child of the importance of safe, responsible behaviour during such excursions.

Signature of parent/guardian:

_____

Date: _____

Permission Letter for Community Field Trips
(Spanish Version)

Estimadas familias:

Durante el año escolar, los alumnos de nuestra clase saldrán para hacer trabajos practicos tanto en la zona de recreo de la escuela como en los alrededores. Estudiaremos temas científicos y descubriremos la importancia de las ciencias en el mundo real. Esperamos hacer este tipo de trabajo como mínimo 10 veces este año.

En primer lugar explicaré el concepto científico y a continuación seguiré la lección con una tertulia y otras actividades en el aula. Por favor comenten estas actividades con su hijo/a. Tambien nos gustaría invitarles a visitar nuestra clase o acompañarnos en estas excursiones.

Gracias.

Por favor, completen y devuelvan este formulario con su hijo/a

Autorizo a mi hijo/a, _____, a salir del recinto escolar para realizar excursiones con la clase, siendo consciente de la posibilidad de riesgo. Le haré saber a mi hijo/a de la importancia de un comportamiento seguro y responsable durante este tipo de excursión.

Firma del padre o la madre o tutor:

_____

Fecha: _____

Permission is granted to reproduce this page for purchaser's class only. Copyright © Trifolium Books Inc.

CHAPTER FOUR Handout 4.6 — **Diversity of Living Things**

Permission Letter for Community Field Trips
(Chinese Version)

亲爱的家长：

本学期我们班学生将要到校园和校园附近的农田参加必多劳动。我们将研究一些科学课题并得出地方次家生活中科学是多么地重要。预计今年我们至少要参加十次这样的活动。首先我解释一下我们的科学字首，接下来在教室里展开讨论。还有更多的一些活动，也许请您家长和您的孩子讨论一下这些活动。我们也很愿意家长送到我们的教室并与我们一道参加这个短途旅行。

谢谢！

请您与您的孩子一起填写并交还如下这一表格：

我同意我的女儿/儿子：_____ 离开校园参加科学游览，我明白会遇到危险，我会提醒我的孩子要注意安全。游览期间，责任自负。

家长鉴字：_____

日期：_____

A Bug's Eye View

- Examine the earthworms with a magnifying glass.
- Draw a picture of what you see inside the magnifying glass on this page.
- Observe the movement of earthworms, then complete the statement below the picture.

Diversity of Living Things Chapter Four Handout 4.7

Three things I noticed about the way the earthworms moved are:

1) _____

_____

2) _____

_____

3) _____

_____

Permission is granted to reproduce this page for purchaser's class only. Copyright © Trifolium Books Inc.

Using the Internet

My advice to teachers: Overcome your fear and apprehension about the internet and give it a try. I promise that you won't regret it. The net is a fun, useful, and extremely powerful tool for both you and your students....

(Dave Sperling quoted in ESL Magazine *1998*
www.eslmag.com)

English language learners, like all students, want to surf the net. The internet can be a frustrating time-waster unless you choose sites and structure activities carefully. Age-appropriate sites may be filled with interesting activities that cause fresh confusions. For example, the Miami Museum of Science site **www.miamisci.org** features "The Atoms Family" with appearances by Dracula, Frankenstein, and the Phantom; in this case, a few words of explanation may overcome the difficulty. In other cases, the "way cool" language is defeating. After you've struggled through the grammatical and cultural explanations required to explain a cute heading like "I ain't nothin' but a rock hound" you'll wish you had never started.

Sites designed for younger children may have an easier reading level but cartoon characters and colouring books are usually considered insulting to older children. You must preview and make judgments about every site you recommend.

Internet research works best when you pre-select the sites.

Searches are likely to be very time-consuming and unproductive for students learning English. For example, the key words "rock cycle" turn up rock gardens, rock music, Plymouth Rock chickens and Rock Hudson. Limiting a search to a single site is an approach that can work.

One internet search project for English language learners might be to find pictures for a class display. The activity "A Class Display on Minerals" is based on suggestions from the School Library Journal web site (Kuntz 2001).

Activity

A Class Display on Minerals

Materials
a computer with internet access
a colour printer
kitchen timer
Handout 4.8: Making a Classroom Display on Minerals from the Internet

Note: Make sure the preferences for **images.google.com** *are set to strict filtering. This is a project that might be done with a small group of students working with a librarian or community volunteer. If you have only one computer connected to the internet, start off with 3 students and guide them through the activity. Then choose one student to explain to the next pair. One student from the new pair remains to assist the next pair.*

- Explain to the students that they will be finding pictures for a bulletin board display.

- Distribute *Handout 4.8: Making a Classroom Display on Minerals from the Internet*.

- Set a fairly short time limit for collecting pictures.

- When students are writing captions for the bulletin board, provide some assistance in proofreading.

- Allow time for students to rehearse explanations for their classmates.

- Organize a gallery walk as described on pages 33 and 34 as a way for all students in the class to present their research.

- If your school has the appropriate facilities, vary this activity by teaching English language learners to prepare Powerpoint demonstrations using only pictures and an occasional title.

Other Ways to Use the Internet

The technological revolution has produced a generation of students who have grown up with multidimensional, interactive media sources, a generation whose understandings and expectations of the world differ from those preceding them. Only by revising educational practices in the light of how our culture has changed can we give these students an appropriate education.

(Heide and Stilborne 1999)

☞ The Earth's crust is an ideal topic for collaborative projects with classes in other parts of the world. Start looking for partners and/or a project at least 6 weeks in advance. The web site **www.globalschoolnet.org** has a projects registry where you can link up with an existing project or seek a class to partner with. Projects such as "Sands of the World" and "Rock Swaps"; may require you to use the postal system as well. This site also has links to samples of student investigations. Some of these are very suitable for English language learners because student writing is less technical than the language used on more scientific web sites.

☞ It would be a wonderful experience if a connection could be made with a class one of your English language learners has left behind. Handle this with sensitivity. Some refugee students might be placing relatives in danger by trying to re-connect with their schools; however, other students may be rushing home after school to exchange gossip with former classmates or may have a relative who is a teacher. In this case, your English language learners would shine as facilitators and their communities could be involved in translation.

☞ When a really good question comes up on your Wonder Wall (see page 44), consider submitting it to "Ask a Geologist" on the Canadian

Community Connection

Families of students may have personal contacts and be able to match your class with a school or a teacher.

Geological Survey web site **www.nrcan.gc.ca/gsc/education_e.html** or the United States Geological Survey web site **walrus.wr.usgs.gov/docs/ask-a-ge.html**.

☞ Try a virtual field trip. The search engine Yahoo has a selection of on-line tours if you follow the Recreation→Travel→Virtual Field Trips links at **www.yahoo.com**. The annual Jason Project investigates such topics as Frozen Worlds (for the 2001–02 school year) and From Shore to Sea (for the 2002–03 school year). The photographs and journals are accessible to English language learners.

☞ Visit a virtual museum such as **www.museumstuff.com** for a selection of displays on rocks, minerals, and geology. Have students learn about volcanos, canyons, and caves using web sites specific for such natural formations near your area.

☞ Arrange a virtual visit to a scientist studying an earthquake, a geologist studying rock formations, or an archeologist excavating a site. Students need to see the connection between science and possible future careers as well as the overlap between science and other subject areas such as history in the case of archeology.

☞ Use search engines for young people such as **www.yahooligans.com** or **www.ask.com** to preview sites that may be current.

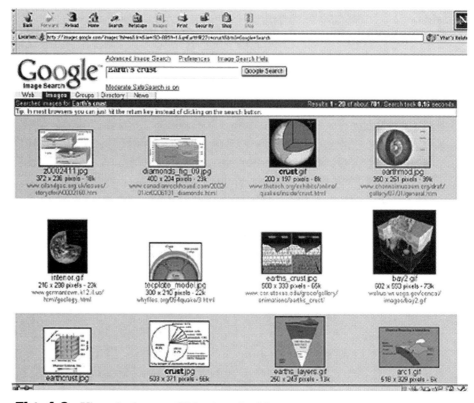

Fig. 4.3 *Visuals from a Web site clarify content*

Making a Classroom Display on Minerals from the Internet

Step 1:
Analyze Your Topic
- Write a word or phrase that describes it.
- Do not use abbreviations.

Step 2:
Decide Where to Search First
- This time go to **images.google.com**. Click one of the image collections or databases.
- Set your timer for 10 minutes.
- Type in one of the target words: broader, narrower, mineral, rock, diamond, mountain, landform, and Mt. Everest.

Step 3:
If You Find Nothing
- Check your spelling.
- Leave out punctuation (for example, rock climbing rather than rock-climbing).

Step 4:
Still Stuck
- If your timer goes off, ask for help.

Step 5:
Select Your Picture
- Choose the illustration you like best.
- Print it.
- Record three interesting facts about it.

Step 6:
Finishing Your Display
- Repeat the steps until you have enough pictures and information to make a display that describes some of the properties of the mineral, how it is mined, what it is used for, and other interesting facts.

The Earth's Crust
CHAPTER FOUR   Handout 4.8

Chapter 5
Assessing Fairly

Using Varied Assessment Techniques

Assessments can be done in many different ways.... They need to be developmentally appropriate, set in contexts familiar to students, and as free from bias as possible.

(National Academy of Sciences 1996)

The standards model of assessment being introduced in schools in Canada and the United States is designed to ensure that all children meet a minimum level of performance based on set criteria.

A three-tiered assessment model is proposed. The first level is called entry-level or diagnostic assessment and is intended to measure the extent of students' existing knowledge or skill to determine whether some of the foundation knowledge needs to be reviewed. The second is progress-monitoring or formative assessment: tasks and mini-tests to help you analyze how your students are learning. This process provides opportunities to give students feedback and to examine your own teaching so that you can move on to new curriculum material or shift gears and re-teach when some students are struggling. The third is summative assessment, consisting of more involved assignments such as research projects and unit tests. This third level is intended for you to evaluate students' performance after they have had a chance to learn new material. At this stage, students must demonstrate, on an individual basis, their understanding of the material through applications and extensions to the core concepts. Marks should be weighted in favour of summative assessments, while marks allotted for progress-monitoring should carry little weight.

This analytical model of assessing is especially helpful to English language learners. Focus on recognizing your students' achievements rather than documenting their failures. This model of assessing also gives students an opportunity to improve their understanding of the curriculum material before a significant proportion of marks are allotted, and for teachers an opportunity to modify teaching strategies to more fairly evaluate their students.

Science lends itself particularly well to assessment that is not based on pencil and paper tasks: observing practical work, creating models, designing prototypes, contributing questions and developing hypotheses. Ironically, students who have grown up in systems in which standardized government examinations determine access to secondary school education often fail to understand the day-to-day nature of North American assessments or the importance we place on the development of both technical and inquiry skills. It is important to be very explicit about how and what you assess.

"And I don't know how teachers mark in Canada. In Korea, the tests take all mark, but here, homework, friendship and many thing so I don't know how to get good mark. The test is not very import."

Marking systems can vary as well. For example, in Japan a correct answer is indicated by an "X" or a perfect circle. An incorrect answer may be indicated by "V." A young Japanese student might confuse your checkmarks with "Vs" and be pleased with a zero. Passing marks may vary from 40% in some schools in Hong Kong to 70% in the Philippines. Remember to explain the marking system used in your school.

It is easy for us to slip into biased assessment in designing tests to see how students can apply content. English language learners may be so unfamiliar with activities like go-cart races or building docks at summer homes that you are no longer letting them show what they have learned.

One type of summative assessment that you can use involves portfolios. Portfolios which include a variety of tasks give students opportunities to show their growth. Here is a portfolio activity that can be applied to any science topic you may be teaching.

Activity

Preparation of a Portfolio

Materials
file folders
Handout 5.1: Checklist of Portfolio Contents

- Decide on a purpose for a portfolio. An assessment portfolio on a unit kept in a file folder is a practical beginning.

- Prepare a checklist of portfolio contents to use as a cover page. An example is provided in *Handout 5.1: Checklist of Portfolio Contents*.

- Explain that for the next month students will be regularly choosing items they are pleased with that demonstrate what they have learned.

- Allot time for student conferences on their portfolios. Small group portfolio discussions are often less threatening and more time efficient than individual conferences. Ask open-ended questions such as: *Which part of your portfolio are you most satisfied with? How could you have done better on this quiz? How would you explain this to me now? What do you think you learned from setting up the display and talking about it? Which of your words can you explain by drawing a picture?*

- Have students explain their portfolios to their parents in first language as part of your reporting procedures. Don't worry too much about the possibility of a student expressing strong feelings about unsatisfactory work; you will notice the parents' reactions and be able to clarify.

Other Ways to Use Varied Assessment Techniques

As teachers, we—like all other teachers—already knew the basics of assessment: observing students closely for signs of growth and mastery, and finding joy in that growth.

(Law and Eckes 1995)

☞ Make it clear from the beginning how you will assess students' work; mark breakdowns and criteria you are looking for on word problems, labs, etc. need to be stated up front. Samples of work at different performance levels help to clarify your expectations.

☞ Brief oral quizzes or one-on-one questioning may reveal a student's knowledge more readily than written work.

☞ Self-evaluation using open-ended prompts for journal writing, such as *Handout 5.2: Sentence Starters*, can give valuable insights into how much students are learning.

☞ If you are using peer assessment in your class, consider looking after the English language learners who immigrated most recently yourself.

☞ If a project is part of a unit, make materials available and arrange for students to have class time to complete the work. A student's grade should not be affected by the family not knowing where to obtain Bristol board or not having the language skills to help with research.

☞ Rehearse the assessments. A mock test or examination using exactly the same format as the assessment tool to be used is extremely helpful, or you can explain that they will be repeating an experiment done in a group.

Checklist of Portfolio Contents

Name: Month and Year:

Contents: Date Completed:

Two examples of science journal entries	1. 2.
Final version of a lab report	
A concept map or Venn diagram	
Photographs of display	
Audio tape describing display	
Short answer quiz	
Self-evaluation checklist of group project	
Top ten words from my personal dictionary	
Summary chart	

Permission is granted to reproduce this page for purchaser's class only. Copyright © Trifolium Books Inc.

Sentence Starters

In your science journal, evaluate your own learning.
Use any of the sentences below as a starting point for your self-evaluation.
How would you improve your learning process?

- This is the first time I ever …
- Before I came to this school, I had learned about …
- Today I learned …
- I want you to know that …
- I used to think … but now I know …
- I am still confused about …
- The hardest part of this work was …
- I am getting better at …
- I'm not sure how to …
- When I did this, I learned …
- Next time I will …
- The next time I work with a partner, I will …
- I want you to notice that …
- This work was frustrating because …
- I was surprised when …
- I wish I had more time …
- I wonder …
- I found out …
- Sometimes I have trouble …
- I am still confused about …
- Three things I know about … are …
- I wish I had asked …
- This question helped me to think about …
- I finally understood when …
- I enjoyed doing this activity because …
- I chose this work for my portfolio because …
- This was an easy assignment for me because …
- This was a hard assignment for me because …
- Compared to other work I have done, this piece is …
- The part I really enjoyed was …

Permission is granted to reproduce this page for purchaser's class only. Copyright © Trifolium Books Inc.

Making Accommodations

Care must be taken in assessing students as they conduct investigations and experiments to differentiate between actual lack of knowledge (or misunderstanding) of science content and physical or linguistic limitations students may have. Reasonable accommodations may be in order for individual students, but the rigor of the learning and assessment challenge needs to be equivalent for all students.

(California Department of Education 2002)

Provincial and state curriculum documents tend to emphasize the necessity of making accommodations but are less likely to be explicit about how this should be done. One of the key principles is to present all students with the same assessment task and to make accommodations in how students perform the task and in your grading. It is important that the whole class understand why English language learners are assessed differently at first.

Imagine that after a few months of Japanese instruction you are required to write a report on continental drift. Your task would be significantly more difficult than that of your Japanese peers, especially if you had not studied geology in a previous science or geography course. How could a teacher make reaching an acceptable standard more possible for you? One way to do this is to ask the other students in the class to imagine themselves in such a situation and write suggestions in their science journals. They may come up with ideas like these:

- I wouldn't want percentage marks until I could get all the answers right.
- Maybe the teacher could show me my mistakes but not take off marks for spelling.
- I wouldn't be able to write fast in Japanese; I'd need more time to draw the characters.
- I would need a dictionary to find the words I know in English but don't know in Japanese.

"True and false are easy. I just make them all false and I usually pass."

Ways to Accommodate English Language Learners

Activity

- Have the other students in the class imagine that they are just learning English.

- Ask them to write down what they think their concerns would be in their science journals.

- Have them comment on how they would want to be evaluated by the teacher and their peers.

- Ask your English language learners to write down their concerns about science class in their science journals.

- Have the English language learners write down what they think is a fair evaluation for them and the other students.

- Have the English language learners and other students compare their comments in small groups.

- Prepare a list of the assessment methods you intend to use, taking into account the student comments. Explain this list clearly to all students so that they know where they stand as far as evaluation.

Other Ways to Make Accommodations

Because ESOL students' test performances do not always reflect their understanding, it is important that content-area teachers not be blinded by one assessment.

(Teemant, Bernhardt and Rodriguez-Munoz 1996)

☞ A school-wide assessment policy for English language learners that is shared with all students is useful. It lets English language learners know what is expected of them so that they have no surprises.

☞ Beginners can sometimes show you what they have learned by writing a test in first language for you to evaluate with the assistance of a bilingual staff member or translator. If a staff member or translator of that language is unavailable, consider having students demonstrate their understanding to you one-on-one. Make sure that the student does not feel that you are critically evaluating them; otherwise you will not get an authentic assessment of their understanding and abilities.

☞ English Language learners who are learning a new alphabet may write very slowly. Giving extra time to write tests may help or you may want to evaluate the amount done. If the student only completed half the paper but that half is substantially correct, use this as the grade for the time being.

- Delay numerical or letter grade assessments for new arrivals until you can record some success. Standards models of assessment require that the achievement recorded is the final summative level rather than diagnostic (entry-level) or formative (progress-monitoring) assessments along the way.

- Phrase questions as clearly as possible. Consider replacing questions phrased as requests with questions beginning with "What," "Where," "When," etc. If you are using word problems, simplify the language as much as possible; avoid making them superfluous or contain unnecessary information.

- Writing a paragraph provides English language learners with scope to show what they know, even if the language they use is far from perfect. Short answer questions are useful.

- Multiple choice assessment items appear easy but the precision of reading required makes them very difficult. Fill-in-the-blanks activities demand a considerable knowledge of English syntax. Although English speakers may not be familiar with the terms noun, verb or adjective, they will likely know that "the *measures* decreases" sounds wrong and "the *temperature* decreases" sounds right; this is a great advantage.

- Summative assessments in science are not the place to evaluate language progress. Concentrate on meaning not on form. The process of taking up a test could include studying model answers and improving answers by re-writing.

- Encourage English language learners to ask for clarification of unusual vocabulary during a test. A quick explanation that a tulip is a flower or rugby is a game can be given. As a general rule, be careful with the wording of questions on a test or exam; be as clear as possible. Encourage students who have limited English language skills to use sketches to explain their answers.

- Consider using varied types of questions on tests and exams, including questions that involve a visual. Labelling diagrams and completing flow charts give English language learners a break from the intensive focus of text-only questions. In addition, using a matching exercise where terms in one column should be paired with a definition in another allows English language learners, who may be unable to express themselves well in English, an opportunity to demonstrate some understanding of scientific terminology.

Glossary of Language Learning Terms

Basic interpersonal communicative skills (BICS): Everyday conversational ability supported by context. This usually takes about two years to acquire.

Closed captions (CC): Subtitles electronically superimposed on the TV screen representing the dialogue and sound effects of a program. Captions are produced by encoding the script of a television program onto the master videotape by computer before broadcast.

Cognitive academic language proficiency (CALP): The language competency required to learn academic subject matter. This is the aspect of language learning that often requires 5–7 years to acquire.

Content area: A specific curriculum area such as science or social studies.

Context embedded: Language for which physical and social clues help clarify the meaning.

Context reduced: Abstract language which requires higher levels of content and language knowledge to be understood.

Intonation: The use of variations in pitch, or voice levels, to convey meaning. For example, the rise in voice level at the end of a statement such as: *You won't be here for the test* turns the statement into a question.

Language acquisition: The unconscious development of language proficiency. Children become fluent in their first language by acquisition. It also plays a role in the development of fluency in a second language.

Language learning: The conscious learning of a language, generally involving a teacher, textbooks, homework, memorization, and drills.

First language (L_1): A student's dominant or primary language.

Second language (L_2): The target language an individual is attempting to learn, in this case English.

Mainstreaming: The process of moving students out of an ESL (English as a Second Language) program of instruction and into content-area courses in which English is the language of instruction.

Non-verbal cues: Information communicated by the use of gestures, eye contact, body movement, time, interpersonal space, pitch, tone, and intonation.

Productive skills: The ability to speak or write a language.

Receptive skills: The ability to understand written and spoken language.

Reading strategies: Methods used in reading to determine the meaning of a text. Some examples are rereading, following text read aloud, using root words to determine the meaning of unfamiliar words, using previous knowledge to determine meaning, predicting content, using graphic organizers, skimming text for information, and recording key points.

Scaffolding: Supports that are initially provided by teachers for students when a new strategy or concept is introduced but are withdrawn over time.

Total physical response (TPR): A language learning activity in which students respond to the verbal commands of the teacher by performing the actions specified.

Word web: A diagram showing the relationship between ideas.

Suggested Web Sites

1. **cnnstudentnews.cnn.com**
   A guide to accompany CNN's closed captioned weekly news programs for students. These programs air in the very early morning. Educators are invited to record them.

2. **images.google.com**
   This web site offers databases of images that students can use when working on certain research projects. Always click preferences boxes for strict filtering

3. **slj.reviewsnews.com**
   Jerry Kuntz' article "Teach and They Shall Find" is available on this web site if you search for it in the Archives link. You can also find it in print in the May 1, 2001 issue of School Library Journal.

4. **starchild.gsfc.nasa.gov**
   This is one of NASA's many web sites, and it offers information about various astronomical topics at two reading levels with an excellent clickable glossary.

5. **walrus.wr.usgs.gov/docs/ask-a-ge.html**
   This is the web site of the United States Geological Survey that offers a similar feature as the Canadian Geological Survey (see web site #17 for more information).

6. **www.ask.com**
   This is the Ask Jeeves web site and is popular with school librarians. It is useful for English language learners because it presents alternate lists of keywords.

7. **www.cellsalive.com**
   This web site has information on plant and animal cells, with associated quizzes, animations and activities.

8. **www.eslcafe.com**
   Dave Sperling, an ESL teacher with extensive teaching experience around the world, created this web site to enable ESL students and teachers to have discussions about ESL topics. There are quizzes and English language tips also available on this site.

9. **www.eslinfusion.oise.utoronto.ca/storiesr.asp**
   Antoinette Gagné, faculty member at the Modern Language Centre at OISE University of Toronto, has a web site which provides information on choosing ESL resources, teaching strategies, assignment ideas, and discussions.

10. **www.fourmilab.ch**
    This web site offers students the ability to view the night sky above the horizon by specifying latitude and longitude using the Your Sky link. The Earth and Moon Viewer link is useful to view various cities around the world as viewed from an Earth satellite in orbit.

11. **www.globalschoolnet.org**
    If you are looking for a way to have students in your class link up with an existing project or you want to find a class to partner with, this web site has a registry of projects that you can choose from.

12. **www.hbschool.com/glossary/math**
    Harcourt School Publishers offers this web site for students having difficulty with mathematical terms.

13. **www.miamisci.org**
    The Miami Museum of Science web site features various activities for students. The language is very trendy so careful explanations are required to prevent confusion for English language learners.

14. **www.museumstuff.com**
    If you want students to visit a virtual museum, this web site provides a selection of displays on a variety of topics from art to history to science.

15. **www.nasm.edu**
    This part of the National Air and Space Museum web site provides a wide variety of virtual hands-on activities related to aviation and space.

16. **www.nrcan.gc.ca/gsc/education_e.html**
    This is the Geological Survey of Canada link on the Natural Resources Canada web site. The feature called "Ask a Geologist" is useful when students are learning about geology. This site also has information on Canadian landscapes and landforms, earthquakes and climate change.

17. **www.pbs.org/teachersource**
    This web site offers teachers lesson plans and information about educational programs on PBS on topics such as art, math and science. Teaching material is tied to the U.S. curriculum.

18. **www.science.edu.sg**
    The Science Centre of Singapore has a ScienceNet section on its web site that features various science questions. The resources section has examples of school projects.

19. **www.smithsonianstore.com**
    This is the web site of the Smithsonian Institute store.

20. **www.cbc.ca/newsinreview**
    This part of CBC's website has an index and ordering information for subscriptions to videotapes and teacher resources.

21. **www.yahoo.com**
    Yahoo is a search engine that has links to various topics. If you follow the Recreation→Travel→Virtual Field Trips link, students can participate In a virtual field trip.

22. **www.yahooligans.com**
    This is another search engine that caters specifically to young people. You will have to find current and relevant topics on here first before having students log on the internet.

23. **www.yourdictionary.com/languages**
    If you are looking for an on-line multilingual dictionary, this web site provides wordlists in 270 languages.

Resources and References

Classroom Resources

1. Bassano, Sharron and Mary Ann Christison. 1992. *Earth and Physical Science: Content and Learning Strategies.* Reading, Massachusetts: Addison-Wesley. ISBN 0-8013-0986-7

   A book designed to prepare English language learners for mainstream science classrooms. The clear readings and illustrations are accompanied by excellent pre-reading and vocabulary activities.

2. Chamot, Ana Uhl, Jim Cummins, Carolyn Kessler, J. Michael O'Malley, and Lily Wong-Filmore. 1997. *ESL: Accelerating English Language Learning.* Glenview, Illinois: Scott, Foresman and Company, 3 books: grade 6, 7, and 8. ISBN 0-673-19682-8

   These three books and the teachers' guides that accompany them are readers for ESL classrooms providing material for integrated units. Force, friction and gravity are the scientific concepts in a chapter called "The Physics of Fun" featuring roller coasters.

3. Kauffman, Dorothy and Gary Apple. 2000. *The Oxford Picture Dictionary for the Content Areas.* New York: Oxford University Press. ISBN 0-19-434-338-3

   A vocabulary teaching tool with many drawings and photographs of background vocabulary. Great as a reference to use when you don't have time to act out or draw a crocodile. A useful source for an ESL teacher preparing students for science topics.

Teacher Resources

1. Chamot, Ana Uhl and J. Michael O'Malley. 1994. *The Calla Handbook: Implementing the Cognitive Academic Language Learning Approach.* New York: Addison-Wesley. ISBN 0-201-53963-2

   Chamot and O'Malley provide examples of how to use current language learning theory, particularly learning strategies, in schools.

2. Coelho, Elizabeth. 1998. *Teaching and Learning in Multicultural Schools: An Integrated Approach, Multilingual Matters.* Clevedon, UK: Clevedon Hall. ISBN 1-85359-383-4

   A practical guide to working in schools with a diverse population. There are useful sections on the impact of background knowledge and on assessment as well as an excellent bibliography.

3. Heide, Ann and Linda Stilborne. 1999. *The Teacher's Guide to the Internet.* Toronto, Ontario: Trifolium Books Inc. ISBN 1-895579-44-9

   This book belongs in your school library. The authors include both classroom activities and clear explanations of internet technologies. Both experts and novices will find it useful.

4. Irujo, Suzanne, ed. 2000. *Integrating the ESL Standards into Classroom Practice: Grades 6–8.* Alexandria, Virginia: Teachers of English to Speakers of Other Languages, Inc. ISBN 0-939791-86-2

   These units for ESL teachers include two with scientific themes: measurement and a study on earthworms.

5. Kessler, Carolyn, ed. 1992. *Cooperative Language Learning: A Teacher's Resource Book.* Englewood Cliffs, New Jersey: Prentice Hall. ISBN 0131736183

   This is a fundamental sourcebook for managing cooperative learning for English language learners. Chapter 4 focuses on cooperative learning in science.

6. Law, Barbara and Mary Eckes. 1990. *The More-than-Just-Surviving Handbook: ESL for Every Classroom Teacher.* Winnipeg, Manitoba: Peguis Publishers Limited. ISBN 0-920541-98-4

   If you can only buy one book, this is it: everything from where to seat students to how to mark their work. A practical guide full of case studies. Passionate and funny as well.

7. ———. 1995. *Assessment and ESL: On the Yellow Big Road to the Withered of Oz.* Winnipeg, Manitoba: Peguis Publishers Limited. ISBN 1-895411-77-7

   All of the contradictions of assessment are included in this practical book for teachers on assessment.

8. Lightbrown, Patsy M. and Nina Spada. 1993. *How Languages Are Learned.* Oxford, UK: Oxford University Press. ISBN 0-19437-00-3

   A slim handbook that is useful as an overview of language-learning issues. Chapter 4 on learner language is especially helpful for content-area teachers.

References

American Association for the Advancement of Science. 1989. *Science for All Americans: A Project 2061 Report on Literacy Goals in Science, Mathematics, and Technology.* Washington, D.C.: American Association for the Advancement of Science. ISBN 0871683415

American Association for the Advancement of Science. 1993. *Benchmarks for Science Literacy: Project 2061.* New York, NY: Oxford University Press

Bassano, Sharron and Mary Ann Christison. 1992. *Earth and Physical Science: Content and Learning Strategies.* Reading, Massachusetts: Addison-Wesley. ISBN 0-8013-0986-7

California Department of Education. 2002. *Draft California Science Framework for K–12 Schools*

Carrasquillo, Angela and Vivian Rodríguez. 2002. *Language Minority Students in the Mainstream Classroom.* Clevedon, UK: Multilingual Matters Ltd. ISBN 1-85359-565-9

Chamot, Ana Uhl, Jim Cummins, Carolyn Kessler, J. Michael O'Malley, and Lily Wong-Filmore. 1997. *ESL: Accelerating English Language Learning.* Glenview, Illinois: Scott, Foresman and Company, 3 books: grade 6, 7, and 8. ISBN 0-673-19682-8

Chamot, Ana Uhl and J. Michael O'Malley. 1994. *The Calla Handbook: Implementing the Cognitive Academic Language Learning Approach.* New York: Addison-Wesley. ISBN 0-201-53963-2

Clemes III, George H. 1998. "Dave Sperling: A Man with a Virtual Passion." *ESL Magazine* (May/June)

Coelho, Elizabeth. 1998. *Teaching and Learning in Multicultural Schools: An Integrated Approach, Multilingual Matters.* Clevedon, UK: Clevedon Hall. ISBN 1-85359-383-4

Collier, V. P. 1987. "Age and Rate of Acquisition of Second Language for Academic Purposes." *TESOL Quarterly* 21, no. 4: 617–41

Common Framework of Science Learning Outcomes: Pan-Canadian Protocol for Collaboration in School Curriculum. 1997. Toronto, Ontario: Council of Ministers of Education. ISBN 0-88987-111-6

Costa, Arthur and Bena Kallick, eds. 2000. *Discovering and Exploring: Habits of Mind, A Developmental Series.* Alexandria, Virginia: Association for Supervision and Curriculum Development. ISBN 0-87120-368-5

Crandall, JoAnne. 1987. *ESL through Content Area Instruction: Mathematics, Science and Social Studies.* Englewood Cliffs, New Jersey: Prentice Hall. ISBN 0132843730

Cummins, Jim. "The Lesson Plan." In *ESL: Accelerating English Language Learning. Teacher's Edition,* by Ana Uhl Chamot, Jim Cummins, Carolyn Kessler, J. Michael O'Malley, and Lily Wong-Filmore (Glenview, Illinois: Scott, Foresman and Company, 1997). ISBN 0-673-19682-8

Early, Margaret. 1990. "Enabling First and Second Language Learners in the Classroom." *Language Arts*, 67 (October)

Echevarria, J., M.E. Voight, and Deborah Short. 2000. *Making Content Comprehensible to English Language Learners: The SIOP Model.* Boston: Allyn & Bacon. ISBN 0-2052-9017-5

Edwards, Viv. 1998. *The Power of Babel: Teaching and Learning in Multilingual Classrooms.* Stoke-on-Trent, UK: Trentham Books Limited. ISBN 1-85856-0950

Faltis, C., and P. Wolf. (Eds.). So much to Say; Adolescents, Bilingualism and ESL in the Secondary School. 1999. New York: Teachers College Press. ISBN 0-8077-3796-8

Fathman, Ann, Mary Ellen Quinn, and Carolyn Kessler. 1992. "Teaching Science to English Learners: Grades 4–8." *National Clearinghouse for Bilingual Education Program Information Guide Series*, no. 11 (Summer)

Fradd, Sandra H., Okhee Lee, Pete Cabrera, Vivian Del Rio, Amelia Leth, Rita Morin, Marisela Ceballos, Maria Santalla, Lucille Cross, and Techeline Mathieu. 1997. "School-University Partnerships to Promote Science with Students Learning English." *TESOL Journal* 7, no. 2: 35–41

Gibbons, Pauline. 1991. *Learning to Learn in a Second Language.* Portsmouth, New Hampshire: Heinemann. ISBN 0-435-08785-1

Heide, Ann and Linda Stilborne. 1999. *The Teacher's Guide to the Internet.* Toronto, Ontario: Trifolium Books Inc. ISBN 1-895579-44-9

Henderson, Jenny and Jerry Wellington. 1998. "Lowering the Language Barrier in the Teaching and Learning of Science." *School Science Review* 79 (March): 288

Irujo, Suzanne, ed. 2000. *Integrating the ESL Standards into Classroom Practice: Grades 6–8.* Alexandria, Virginia: Teachers of English to Speakers of Other Languages, Inc. ISBN 0-939791-86-2

Kauffman, Dorothy and Gary Apple. 2000. *The Oxford Picture Dictionary for the Content Areas.* New York: Oxford University Press. ISBN 0-19-434-338-3

Kessler, Carolyn, ed. 1992. *Cooperative Language Learning: A Teacher's Resource Book.* Englewood Cliffs, New Jersey: Prentice Hall. ISBN 0131736183

Kuntz, Jerry. 2001. "Teach and They Shall Find." *School Library Journal* (May 1). Available on: slj.reviewsnews.com

Laufer, B. "How Much Lexis Is Necessary for Reading Comprehension?" In *Vocabulary and Applied Linguistics*, edited by H. Béjoint and P. Arnaud (London, UK: Macmillan, 1992, 126–32)

Law, Barbara and Mary Eckes. 1990. *The More-than-Just-Surviving Handbook: ESL for Every Classroom Teacher*. Winnipeg, Manitoba: Peguis Publishers Limited. ISBN 0-920541-98-4

———. 1995. Assessment and ESL: *On the Yellow Big Road to the Withered of Oz*. Winnipeg, Manitoba: Peguis Publishers Limited. ISBN 1-895411-77-7

Lee, O. and S. H. Fradd. 1996. "Literacy Skills in Science Performance among Culturally and Linguistically Diverse Students." *Science Education* 80: 651–71

Lewis, Marilyn. 1997. *Journeys in Language and Learning: ESOL Students in Elementary Classrooms around the World*. Scarborough, Ontario: Nelson Canada. ISBN 0-17-695545-2

Lightbrown, Patsy M. and Nina Spada. 1993. *How Languages Are Learned*. Oxford, UK: Oxford University Press. ISBN 0-19437-00-3

Lorenzo, Mercedes, Brian Hand, Vaughan Prian. 2001. "Writing for Learning in Science: Producing a Video Script on Light." *School Science Review* 82, no. 301 (June): 33–9

McKeon, D. 1994. "When Meeting Common Standards Is Uncommonly Difficult." *Educational Leadership* 51, no. 8: 45–9

Ministry of Education and Training. 1999. *The Ontario Curriculum Grades 9–12. English as a Second Language and English Literacy Development*. Ontario: Queen's Printer for Ontario. ISBN 0-7778-8338-4

National Academy of Sciences. 1996. *National Science Education Standards*. Washington, D.C.: National Academy Press. ISBN 0-309-05-326-9

O'Malley, J. Michael. 1996. *Authentic Assessment for English Language Learners*. Reading, Massachusetts: Addison-Wesley. ISBN 0-2-1159151-0

Ovando, Carlos J. and Virginia P. Collier. 1998. *Bilingual and ESL Classrooms: Teaching in Multicultural Contexts*. Boston: McGraw-Hill. ISBN 0-07-047959-3

Prophet, Bob and Peter Towse. 1999. "Pupils' Understanding of Some Non-Technical Words in Science." *School Science Review* 81, no. 295 (December): 79–81

Rigg, Pat and Virginia G. Allen. 1989. *When They Don't All Speak English: Integrating the ESL Student into the Regular Classroom*. Urbana, Illinois: National Council of Teachers of English. ISBN 0-8141-5693-2

Rosenthal, Judith W. 1996. *Teaching Science to Language Minority Students, Multilingual Matters.* Clevedon, UK: Clevedon Hall. ISBN 1-85359-272-2

Sacks, Oliver. 2001. *Uncle Tungsten: Memories of a Chemical Boyhood.* Toronto, Ontario: Alfred A. Knopf Canada. ISBN 0-676-97537-2

Saville-Troike, M. 1984. "What Really Matters in Second Language Learning for Academic Achievement." *TESOL Quarterly* 18, no. 2: 199–219

Scarcella, Robin and Rebecca Oxford. 1992. *The Tapestry of Language Learning: The Individual in the Communicative Classroom.* Boston: Heinle & Heinle Publishers. ISBN 0-83842359-0

Scarcella, Robin. 1999. *Teaching Language Minority Students in the Multi-Cultural Classroom.* Englewood Cliffs, New Jersey: Prentice Hall. ISBN 0-138518254

Scott, JoAnna. 1993. *Science and Language Links: Classroom Implications.* Portsmouth, New Hampshire: Heinemann. ISBN 0-435-08338

Szeto, Sandy. 2002. *Take an Ecowalk to Explore Science Concepts.* Toronto, Ontario: Trifolium Books Inc. ISBN 1-55244-027-03

Szeto, Sandy and Catherine Little. 2001. *Take a Mathwalk to Learn about Mathematics in Your Community.* Toronto, Ontario: Trifolium Books Inc. ISBN 1-55244-009-5

Tang, Gloria. 1997. "Pocket Electronic Dictionaries for Second Language Learning: Help or Hindrance." *TESL Canada Journal* 15, no. 1: 39–57

Teemant, Annela, Elizabeth Bernhardt and Marisol Rodriguez-Munoz. 1996. "Collaborating with Content-Area Teachers: What We Need to Share." *TESOL Journal* (Summer): 16–20

Verplaetse, Lorrie Stoops. 1998. "How Content Teachers Interact with English Language Learners." *TESOL Journal* 7, no. 5

Williams, Peter and Saryl Jacobson. 1997. *Take a Technowalk to Learn about Materials and Structures.* Toronto, Ontario: Trifolium Books Inc. ISBN 1-895579-76-7

———. 2000. *Take a Technowalk to Learn about Mechanisms and Energy.* Toronto, Ontario: Trifolium Books Inc. ISBN 1-55244-004-4

The Authors

Catherine Little is an Instructional Leader with a large school board in Ontario, Canada. She is an experienced Mathematics and Science & Technology Teacher at the middle school level. Her writing has appeared in such journals as *The Crucible* (Science Teachers' Association of Ontario), *Mathematics teaching in the Middle School* (National Council of Teachers of Mathematics) and *Educational Leadership* (Association for Supervision and Curriculum Development. She has worked extensively on textbook development and resource materials for teachers. She is a recipient of the York University Faculty of Education alumni Association Teacher of the Year Award and has coached a provincial level championship Science Olympiad Team.

Jane Hill and **Jane Sims** have both been involved with language teaching as English teachers and as English as a Second Language department heads in secondary schools in Toronto. They have been frequent presenters at conferences such as TESL Ontario and TESOL. They worked together on cross-curricular units of study entitled, *One Earth* and on course profiles for the provincial Ministry of Education and Training. They met Catherine Little on a writing team preparing a Resource Guide for grade nine ESL students in the Toronto District School Board. This is their first book together.